ISBN 978-0-266-88200-8
PIBN 10903981

SPECIAL PUBLICATION 475

U.S. DEPARTMENT OF COMMERCE / National Bureau of Standards

The Electron Factor in Catalysis on Metals

NATIONAL BUREAU OF STANDARDS

The National Bureau of Standards[1] was established by an act of Congress March 3, 1901. The Bureau's overall goal is to strengthen and advance the Nation's science and technology and facilitate their effective application for public benefit. To this end, the Bureau conducts research and provides: (1) a basis for the Nation's physical measurement system, (2) scientific and technological services for industry and government, (3) a technical basis for equity in trade, and (4) technical services to promote public safety. The Bureau consists of the Institute for Basic Standards, the Institute for Materials Research, the Institute for Applied Technology, the Institute for Computer Sciences and Technology, the Office for Information Programs, and the Office of Experimental Technology Incentives Program.

THE INSTITUTE FOR BASIC STANDARDS provides the central basis within the United States of a complete and consistent system of physical measurement; coordinates that system with measurement systems of other nations; and furnishes essential services leading to accurate and uniform physical measurements throughout the Nation's scientific community, industry, and commerce. The Institute consists of the Office of Measurement Services, and the following center and divisions:

Applied Mathematics — Electricity — Mechanics — Heat — Optical Physics — Center for Radiation Research — Laboratory Astrophysics[2] — Cryogenics[2] — Electromagnetics[2] — Time and Frequency[2].

THE INSTITUTE FOR MATERIALS RESEARCH conducts materials research leading to improved methods of measurement, standards, and data on the properties of well-characterized materials needed by industry, commerce, educational institutions, and Government; provides advisory and research services to other Government agencies; and develops, produces, and distributes standard reference materials. The Institute consists of the Office of Standard Reference Materials, the Office of Air and Water Measurement, and the following divisions:

Analytical Chemistry — Polymers — Metallurgy — Inorganic Materials — Reactor Radiation — Physical Chemistry.

THE INSTITUTE FOR APPLIED TECHNOLOGY provides technical services developing and promoting the use of available technology; cooperates with public and private organizations in developing technological standards, codes, and test methods; and provides technical advice services, and information to Government agencies and the public. The Institute consists of the following divisions and centers:

Standards Application and Analysis — Electronic Technology — Center for Consumer Product Technology: Product Systems Analysis; Product Engineering — Center for Building Technology: Structures, Materials, and Safety; Building Environment; Technical Evaluation and Application — Center for Fire Research: Fire Science; Fire Safety Engineering.

THE INSTITUTE FOR COMPUTER SCIENCES AND TECHNOLOGY conducts research and provides technical services designed to aid Government agencies in improving cost effectiveness in the conduct of their programs through the selection, acquisition, and effective utilization of automatic data processing equipment; and serves as the principal focus wthin the executive branch for the development of Federal standards for automatic data processing equipment, techniques, and computer languages. The Institute consist of the following divisions:

Computer Services — Systems and Software — Computer Systems Engineering — Information Technology.

THE OFFICE OF EXPERIMENTAL TECHNOLOGY INCENTIVES PROGRAM seeks to affect public policy and process to facilitate technological change in the private sector by examining and experimenting with Government policies and practices in order to identify and remove Government-related barriers and to correct inherent market imperfections that impede the innovation process.

THE OFFICE FOR INFORMATION PROGRAMS promotes optimum dissemination and accessibility of scientific information generated within NBS; promotes the development of the National Standard Reference Data System and a system of information analysis centers dealing with the broader aspects of the National Measurement System; provides appropriate services to ensure that the NBS staff has optimum accessibility to the scientific information of the world. The Office consists of the following organizational units:

Office of Standard Reference Data — Office of Information Activities — Office of Technical Publications — Library — Office of International Standards — Office of International Relations.

---

[1] Headquarters and Laboratories at Gaithersburg, Maryland, unless otherwise noted; mailing address Washington, D.C. 20234.
[2] Located at Boulder, Colorado 80302.

The Electron Factor in Catalysis on Metals

Proceedings of a workshop held at the
National Bureau of Standards
Gaithersburg, Maryland, December 8-9, 1975

Edited by

L. H. Bennett

Institute for Materials Research
National Bureau of Standards
Washington D.C. 20234

Sponsored by

National Bureau of Standards

National Science Foundation

Energy Research and Development Administration

U.S. DEPARTMENT OF COMMERCE, Juanita M. Kreps, Secretary

Dr. Betsy Ancker-Johnson, Assistant Secretary for Science and Technology

NATIONAL BUREAU OF STANDARDS, Ernest Ambler, Acting Director

Issued April 1977

Library of Congress Catalog Card Number: 77-600009

National Bureau of Standards Special Publication 475

Nat. Bur. Stand. (U.S.), Spec. Publ. 475, 217 pages (Apr. 1977)

CODEN: XNBSAV

U.S. GOVERNMENT PRINTING OFFICE
WASHINGTON: 1977

For sale by the Superintendent of Documents, U.S. Government Printing Office
Washington, D.C- 20402 - Price $2.80

Stock No. 003-003-01764-7

GENERAL ABSTRACT

This book presents the proceedings of a Workshop on the Electron Factor in Catalysis on Metals held at the National Bureau of Standards, Gaithersburg, Maryland on December 8-9, 1975. The Workshop was sponsored by the Institute for Materials Research, NBS, the Division of Materials Research of the National Science Foundation, and the Division of Conservation Research and Technology of the Energy Research and Development Administration. The purpose of the Workshop was to review the most recent experimental and theoretical investigations on catalysis on metals and related topics, and to bring together chemists, chemical engineers, surface scientists, and solid state physicists and chemists involved in research related to this topic. These proceedings summarize the four panel sessions into which the Workshop was organized: Experimental Techniques, Effect of Alloying, Geometrical Effects, and Electronic Structure.

KEY WORDS: Catalysis; Characterization; Chemisorption; Electronic factor; Geometric factor; Metals; Surfaces.

Welcome to the Workshop on the Electron Factor in Catalysis on Metals

E. Ambler

We are pleased to be joining the National Science Foundation and the Energy Research and Development Administration in the sponsorship of this workshop. As the Acting Director of the National Bureau of Standards, it is my pleasant duty to officially welcome you this morning.

As some of you may know, this workshop is being held on the eve of the Bureau's 75th anniversary. The National Bureau of Standards was founded by Congressional legislation in 1901. Although that original legislation has been amended several times, three themes have persisted that reflect the scope of NBS activities. First, it is our responsibility to provide the nation's standards of measurement, second, we determine physical constants and properties of materials, and finally, we provide other agency support. Approximately 3,500 employees work at our site here in Gaithersburg, our site in Boulder, Colorado and our radio stations in Fort Collins, Colorado and the Hawaiian Islands. Our activities range from maintaining the nation's primary frequency standard and determining precise atomic constants and data, to carrying out fire research and investigating some aspects of the process of catalysis. That is to say from making precise measurements on very well defined systems to attempting to unravel processes that are inherently very complex.

When I joined the Bureau nearly 25 years ago as a bench level physicist, the process was of great economic value to this nation's commerce, particularly to the oil industry. Much has happened in the intervening 25 years enabling us to better understand the basic nature of the catalytic process.

Moreover, catalysis remains not only a significant factor in our industrial economy, it has become of even greater importance as we strive to clean up our environment and move toward energy independence. There is an increasing need to invent less expensive catalysts and to find ways of extending life of those now in use.

Science has made great progress in this field over the same time period. Twenty years ago we could only speculate on the role the electron factor might play in the catalytic process. Today, we are able to investigate in some detail the role played by the valence electrons of the various constituents involved in a reaction. Over the past few years, meaningful considerations of the electron factor in catalysis have been made more detailed and specific largely because of the new techniques developed, especially within the surface science community.

I have always felt that a sign of vigor in a given area of science and technology is when theorists and practitioners stand side by side and work on the same problems. This is the case with catalysis. This workshop offers an opportunity for workers from many fields--from theoretical physicists to surface chemists to catalytic engineers--to broaden and strengthen the common thread of communication. Hopefully, this type of interaction will send us back to our respective home bases with increased enthusiasm and renewed dedication.

My best wishes for a most successful workshop.

Overview of the Workshop

L. H. Bennett

Traditionally, the term electron factor in catalysis has denoted a
relationship between catalytic activity and the bulk electronic structure of the catalyst. While there is abundant empirical evidence of the
importance of the electron factor in heterogeneous catalysis, its precise
definition and relation to other factors, particularly the geometric
factor, have yet to be established. The interdependence of electronic
and geometric factors in bulk bonding is well known: examples include
the Hume-Rothery rules of alloy phase stability, relating valence, size,
and electronegativity; and Pauling's famous equation relating bond
length to bond strength. Chemical processes at a solid-fluid interface
have not, thus far, been similarly systematized. However, many new and
powerful techniques for the analysis and characterization of clean
surfaces and small particles, and of chemical complexes on such surfaces
have been developed in recent years. It was the aim of this workshop to
move toward systematic understanding of this problem through interdisciplinary action. Specific panel topics were: Experimental Techniques, the Effects of Alloying, Geometrical Effects, and the Theory of
Electronic Structure.

The workshop brought together workers in a range of fields including
theoretical physicists engaged in band structure and surface calculations,

theoretical chemists doing chemical bonding calculations, surface chemists working on clean surfaces and single crystals, practical catalytic chemists engaged in small particle and support effect studies, electrochemists concerned with activity in aqueous environments, etc.

While progress toward quantitative expression of the Electron Factor in catalysis was limited, this aim served as a focus for the clarification of the capabilities and accomplishments of many experimental and computational techniques. The interdisciplinary emphasis of the conference was one of its most fortunate aspects; many of the distinguished experts present found clarification of the relation between their specialties and catalysis.

Keynote Address

Catalysis by Metals:

Concepts, Factors and Reactions

M. Boudart

Department of Chemical Engineering

Stanford University

Stanford, California 94305

INTRODUCTION

In the past few years, there has been a growing number of studies overlapping the physics and chemistry of metal surfaces. These have generated inquiries into the physical basis of heterogeneous catalysis (1) or the relationship between so-called surface science and catalysis (2) or the impact of the new physical tools on surface chemistry (3). In spite of the limited thrust of these recent studies, confined as they are to metals which constitute only a very small fraction of catalytic materials, the new results have already changed many of the traditional concepts of heterogeneous catalysis. Although these results obtained often at very low pressures on large chunks of metals may not seem at first to be relevant to catalytic reactions run at high pressures on metallic clusters containing about one hundred atoms, closer examination shows how concepts and even reaction mechanisms can be transposed from low to high pressures (4) or from single crystals to small clusters (5). The purpose of this introduction is to survey some of the new emerging concepts or some modifications in the old concepts in heterogeneous catalysis by metals and alloys.

CONCEPTS

The catalytic reaction engineer (6) uses a number of concepts evolved between about 1920 and 1960 in what might be called the Langmuirian period of surface science. In the post-Langmuirian period, new concepts have emerged which will be taken up by the next generation of practitioners of heterogeneous catalysis. As happens frequently in science, the new concept does not supplant the older one but extends it to new situations which were not envisaged in the past.

Thus, the Langmuirian idea of adsorption site remains valid of course but recent work shows that off-site adsorption is important not only in the case of physisorption but also in chemisorbed compressed layers, for instance of carbon monoxide on palladium (7). This is an important new idea, as carbon monoxide is frequently used to titrate metallic sites and also because

6

carbon monoxide and hydrogen are the basis of many catalytic processes toward the synthesis of fuels and chemicals. Besides, off-site chemisorption may be a widespread phenomenon. In fact, since the compressed layer forms a coincidence lattice over the metal lattice, off-site chemisorption is related to corrosive chemisorption or surface reconstruction where the adsorbate forms with the outer layer of metal atoms a coincidence lattice over the subjacent metal lattice, as for instance in the case of sulfur over low index faces of copper (8). This is very different from the Langmuirian pictures of immutable checkerboards. Corrosive chemisorption is the first step toward destructive aging of a metallic catalyst.

Another fundamental idea of Langmuir was the monolayer due to saturation of the adsorption sites. Many examples are now known of ordered surface structures consisting of definite fractions of a monolayer. As many chemisorbed molecules are used to titrate metallic surface sites, a knowledge of these non-classical stoichiometries and of the conditions under which they are obtained, is very desirable in applied catalysis. Even more important perhaps is the reason behind these ordered chemisorbed sub-monolayers, namely attractive forces between chemisorbed species on metals. While repulsive forces between chemisorbed species have been discussed many times during the Langmuirian periods, attractive forces between adsorbates were recognized only in the case of physisorption. Quite recently, attraction between chemisorbed species on metals has been discovered not only experimentally but also theoretically (9). The role of these interactions in catalytic reactions remains to be explored.

Another Langmuirian postulate was that a molecule striking a site would rebound if that site were already occupied. It is now clear that this site exclusion principle is violated in many instances, at least on clean metal surfaces, where the striking molecule may be held at the surface in a weakly bound precursor state long enough so that it can diffuse to the site where adsorption takes place (10). These precursor states must be of great importance

in catalytic reactions. They seem to be bound to the metal surfaces with energies intermediate between those found in physisorption and chemisorption. A striking example is
undissociated methane on tungsten (100) with a binding energy of more than 28 kJmol^{-1} (11).

Another advance of the new era is the identification of adsorption sites with defined
surface atoms. This is still a particularly controversial area but rapid progress is being made (12).
An early example is Estrup's assignment of hydrogen atom chemisorption to bridging positions
between any pair of tungsten atoms at a W(100) surface (13). Thus for the first time, it is now
possible to treat adsorbate-metal complexes with almost normal ideas of molecular structure
and bonding.

As to kinetics of desorption, Langmuir first proposed, besides a rate proportional to
fraction of surface covered θ, a rate decreasing exponentially with θ, the first of a long series
of phenomenological treatments of adsorption, desorption and catalytic reactions to which the
names of Temkin, Zeldovich and Wagner are attached in particular (14). Quite novel on the
contrary is the finding by Madix of a self-accelerating (autocatalytic or explosive) rate of
desorption from a Ni(110) surface on which formic acid has been preadsorbed (15). This is the
first new kinetic pattern of surface reactivity in forty years and is a remarkable example of
non-classical behavior.

Finally, consider the two modes of catalytic reactions between molecules A and B
either involving reaction between both A and B in chemisorbed states (Langmuir-Hinshelwood)
or involving reaction between non-chemisorbed B with pre-chemisorbed A (Rideal-Eley). The
distinction between these modes of reaction has generated byzantine discussions over the past
thirty years. Yet, today, a decision between them can be reached from molecular beam
reactive scattering studies now performed in a number of post-Langmuirian laboratories.

In summary, off-site chemisorption, corrosive chemisorption, fractional ordered
monolayers, attraction between chemisorbed species, precursor states, adsorption sites with

assigned structures, explosive surface desorption, molecular beam scattering at surfaces, are just a few examples that have already changed profoundly our Langmuirian ideas of looking at catalytic reactions at metallic surfaces. The new findings should contribute to the clarification of the traditional factors that were formulated during the Langmuirian period of catalysis: the electronic and geometric factors.

FACTORS

The old distinction in catalysis by metals between geometric and electronic effects is confusing and retains by now only historical interest.

Instead of geometric effect, it is better to talk about effect of structure, the latter being defined by the distribution of surface atoms of given coordination numbers, as varied by exposing various crystallographic planes or changing particle size in the critical range between 1 and 10 nm (16). Structure is different from geometry. Indeed, another post-Langmuirian example consists of hexagonal overlayers on the (100) planes of iridium, platinum and gold (17,18). But although the geometry of the atoms in these overlayers is the same as that found on the (111) faces of these metals, the structure, as defined here, is different in both cases (19). Besides, the ultimate difference between atoms of different geometry or structure, must be ascribed in last analysis to electronic effects, as pointed out a long time ago (20).

Thus, it is best to avoid not only the geometric factor but also the electronic factor as the latter also denotes reasonings associated with the now discarded rigid band theory of alloys. Perhaps the best alternative to electronic effect would be ligand effect introduced by Sachtler and his school to denote the change in reactivity of a metallic atom A when some or all of its A neighbors are replaced by atoms of metal B. Similarly then, the difference in reactivity between two surfaces of pure metals A and B of identical structure, would be ascribed to the different ligand factors of A and B.

9

Another factor related to the ill-defined geometric effect is that which is related to the need for more than one surface atom in the rate determining process of the catalytic reaction. What is needed perhaps is a multiple site, or multiplet, or ensemble, the properties of which may be affected by either or both of the structural and ligand factors as defined here. To determine a priori the relative importance of these factors in catalysis by metals and alloys, ultimately in a quantitative manner, remains a formidable challenge in our post-Langmuirian era of surface science.

REACTIONS

What can be done today is to survey the empirical evidence in order to rank the relative importance of structural and ligand factors for various catalytic reactions in the hope of achieving classifications which suggest future work of a fundamental nature.

Such a survey has been conducted (21) and though the list of reactions is thus far limited, it is noteworthy that ten years ago, this simple task would have been impossible for lack of data and even five years ago, it would have led to an ever much more restricted list of reactions.

In essence, what has been found for reactions performed in excess hydrogen on metals of Group VIII and alloys between metals of Groups VIII and Ib is that two classes of reactions emerge. In the first class are reactions involving breaking or making of H-H bonds and C-H bonds, e.g. hydrogenation of alkenes and aromatics. In the second class are reactions where C-C bonds are broken as in hydrogenolysis or cracking as well as ammonia synthesis or decomposition where N-N bonds are broken or made.

Reactions of the first class are found to be structure insensitive while reactions of the second class are structure sensitive. When metals of Group VIII are surveyed for their catalytic activity, it is found that rates for reactions of Class I vary much less than rates for

10

reactions of Class II. Finally, in studies of the effect of adding Group Ib atoms to Group VIII metals, this latter effect is found to be much more important for Class II than for Class I reactions.

To explain these findings may not be as formidable a task as predictions of catalytic specificity of metals. Even so all that can be done today with the available classification is to speculate. One possibility is that in reactions of Class I, a single surface atom or maybe at the most a pair of them is required in the rate determining process of the catalytic reaction. By contrast in reactions of Class II, a multiple site might be required. To check this simple hypothesis may not be an impossible or remote task in the age of post-Langmuirian surface science.

REFERENCES

1.   The Physical Basis for Heterogeneous Catalysis, E. Drauglis and R. I. Jaffee, Eds. Plenum Press, New York, 1975.

2.   M. Boudart, Chem. Tech. 4, 748 (1974).

3.   J. T. Yates, C & E News, August 26, 1974.

4.   M. Boudart, D. M. Collins, F. V. Hanson and W. E. Spicer, J. Vac. Sci. Technology, in press.

5.   M. Boudart, in Interactions on Metal Surfaces, R. Gomer Ed., Springer-Verlag, Berlin Heidelberg 1975, p. 286.

6.   J. J. Carberry, Chemical and Catalytic Reaction Engineering, McGraw Hill, New York 1976.

7.   J. C. Tracy and P. W. Palmberg, J. Chem. Phys., 51, 4852 (1969).

8.   J. Bénard, Catal. Rev. 3, 93 (1970).

9.   T. L. Einstein and J. R. Schrieffer, J. Vac. Sci. Technol. 9, 956 (1972).

10. G. Ehrlich, J. Chem. Phys. 59, 473 (1955).

11. J. T. Yates and T. E. Madey, Surface Sci. 28, 437 (1971).

12. T. N. Rhodin and D. L. Adams, Treatise on Solid State Chemistry, B. Hannay, Ed., Vol 6A, Chap. 5, in press.

13. P. J. Estrup and J. Anderson, J. Chem. Phys. 45, 2254 (1966).

14. M. Boudart, in Treatise of Physical Chemistry, H. Eyring, W. Jost and Henderson, Eds., Vol. 7, Chap. 7, Academic Press, Inc., New York 1975.

15. J. L. Falconer, J. G. McCarty and R. J. Madix, Surface Sci. 42, 329 (1974).

16. R. Van Hardeveld and F. Hartog, Surface Sci. 15, 189 (1969).

17. T. N. Rhodin, P. W. Palmberg and E. W. Plummer, in The Structure and Chemistry of Solid Surfaces, G. A. Somorjai, Ed., Wiley, New York 1969, paper No. 22.

18. T. A. Clarke, I. D. Gray and R. Mason, Surface Sci. 50, 137 (1975).

19. C. R. Helms, personal communication.

20. M. Boudart, J. Am. Chem. Soc., 72, 1040 (1950).

21. M. Boudart, Proc. VIth Intern. Congr. Catal., Pitman Press, Bath, in press.

EXPERIMENTAL
TECHNIQUES

Moderator:

G.HALLER

Yale University

Lecture by:

J.T.YATES

NBS

Session 1.

Experimental Techniques

Panel Members:

J. Katzer,
Univ. of Del.

R. Park,
Univ. of Md.

T. Rhodin,
Cornell Univ.

J. T. Yates,
NBS

Recorder:

T. E. Madey,
NBS

EXPERIMENTAL METHODS IN HETEROGENEOUS CATALYSIS

J. T. Yates, Jr.
Surface Processes and Catalysis Section
Institute for Materials Research
National Bureau of Standards
Washington, D.C. 20234

I. Introduction

Research in the field of heterogeneous catalysis currently involves the use of many types of measurement techniques. Ultimately one wishes to employ these measurement techniques to design catalysts which enhance the rates and selectivity of catalytic processes as well as the useful lifetime of the catalyst. Along the way to achieving these very practical objectives, we have the exciting intellectual possibility of understanding the atomic and electronic features of complex chemical processes which are specifically promoted on the surface of solid catalysts.

Ideally, fundamental research on a catalytic process should give us definitive information of three kinds:

1. Character and surface concentration of active sites.

2. Identity of catalytic intermediates and mechanism of the reaction.

3. Rate of the catalytic reaction.

In practice, we must often be satisfied with less than this complete picture of the catalytic reaction – in many cases this is because of our limited ability to make the necessary physical or chemical measurements. This is particularly true in the case of catalytic intermediate identification since spectroscopic techniques often detect the major adsorbed species which may not be catalytic intermediates.

In this short paper I want to give a brief summary of some of the best examples of current catalytic research which illustrate the state of the art in measuring definitive features of the three types listed above. A longer

15

review is available covering other topics related to this subject.[1]

II. Measurement of the Character and Surface Concentration of Active
    Sites using Electron Spin Resonance

Electron Spin resonance (ESR) techniques are especially well suited for quantitative studies of the nature of active sites on catalysts. The method is exceedingly sensitive, and when applied to high surface area solids, is capable of detection of $\lesssim 10^5$ spins/cm^2. In recent work by Boudart and coworkers[2], the active site on MgO which catalyzes H_2/D_2 exchange has been characterized, using ESR. It has been found that approximately one site in 10^6 is active for exchange. The site is postulated to be an array of O^- radical ions associated with a nearby hydroxyl group which participates in exchange. There is a remarkable correlation of the catalytic rate of H_2/D_2 exchange with the concentration of active sites measured, over a range of several orders of magnitude. By appropriate surface treatment, the active site may be caused to appear and disappear and a concomitant variation in catalytic activity is observed.

In another thorough ESR investigation, Voorhoeve and coworkers[3] have studied the ESR spectrum of WS_2 catalysts. The objective was to determine the nature of active sites for benzene hydrogenation as well as for other reactions. Again, it was found that catalytic activity was associated with surface defect sites where W^{+3} ions are present in contrast to W^{+4} ions in the normal crystal lattice. The W^{+3} sites exhibit lower coordination with neighboring S^{-2} ions than is found in the case of bulk W^{+4} ions. These sites are preferentially located at the edges of WS_2 crystallites; enhancement of their concentration by various methods including preparation of sulphur deficient non-stoichiometric WS_2 leads to increased catalytic activity which is proportional to the W^{+3}

16

ESR signal over a range of more than 3 orders of magnitude as shown in
Fig. 1. It is postulated that W^{+3} preferentially binds benzene via a
π-complex interaction and that catalytic hydrogenation of these bound
species occurs.

III. **Measurement of the Structure of Adsorbed Species and Catalytic**
     Intermediates using IR Spectroscopy.

Infrared (IR) spectroscopy has become a widely used technique for the
study of heterogeneous catalysis.[4,5] In general, it has been found that
the principles employed in interpreting IR spectra of chemical compounds
are also useful in studies of adsorbed species. Thus, the concept of group
vibrational frequencies associated with different functional groups seems
to be valid on surfaces just as in molecules. Also, the belief that electronic
effects cause small shifts in group vibrational frequencies is widely accepted.

IR spectroscopy has been applied to the study of the mode of adsorption
of hydrogen by ZnO catalysts.[6,7] Through the use of HD as an adsorbate,
it was found that hydrogen chemisorption occurs on a ZnO pair site, with
preferential adsorption in the
$$
\begin{array}{cc} & \text{D...H} \\ | & | \\ \text{Zn} & \text{O} \end{array}
$$
conformation at low temperatures.
Heating to ∿ 300 K causes an irreversible change of the DH conformation to
$$
\begin{array}{cc} \text{H...D} & \\ | & | \\ \text{Zn} & \text{O} \end{array}
$$
as shown by the IR intensity behavior in Fig. 2. Control experiments
involving D_2 or H_2 adsorption indicate that preferential Zn-D or O-H adsorption
does not occur for these homomolecular adsorbates.[8]

Kinetic and thermodynamic isotope effects are postulated to be responsible for this unusual behavior. Irrespective of the explanation of
the effect, it is necessary to conclude that H_2 adsorbs on a ZnO pair site.
Thus, in this example, it is seen that IR is a powerful tool for ascertaining
the general bonding nature of an adsorbed species.

17

A second example of the utility of IR spectroscopy for structural studies of adsorbed species has to do with CO chemisorption by transition metals. It is well known that two general kinds of CO are often observed to adsorb on transition metals - a form with CO stretching frequency in the range 2000-2100 cm^{-1} [linear-CO, sp hybridized] and a form exhibiting peak adsorbance below 2000 cm^{-1} [bridged-CO, sp^2 hybridized]. There has long been a controversy about the assignment of the "bridged-CO species". In a recent study, Sachtler and coworkers[9] investigated the spectrum of CO on a range of Pd/Ag alloys. On pure Pd, the major infrared band is the "bridged band". As Ag is alloyed with Pd, there is a diminution of the bridged band and an increase in intensity of the linear band as shown in Fig. 3. This is interpreted as being due to the statistical reduction of Pd_2 sites due to Ag alloying. A secondary feature of the experiments is also of importance. Over a wide range of Pd/Ag alloy compositions, the two CO bands are observed to exhibit almost constant frequencies. This behavior may be ascribed to the electronically independent nature of the constituent atoms in the alloy. This conclusion has been confirmed by recent x-ray photoelectron spectroscopy (ESCA) valence band studies of Pd/Ag alloys where it is seen that the d-electrons from each atom are behaving atomically, in contrast to the predictions of the rigid band model for alloys.[10]

IV. Measurement of the Electronic Character of Adsorbed Species Using Ultraviolet Photoelectron Spectroscopy (UPS)

The use of monochromatic ultraviolet light to eject valence level photoelectrons from solids and from surfaces containing adsorbed species has recently become widespread in surface physics and chemistry. For a metal, the highest energy photoelectrons will be ejected from the top of the conduction band; electrons of lower kinetic energy will be generated in an energy distribution curve due to photoemission throughout the band to a depth of $(h\nu-\phi)$, where ϕ is the work function of the surface. Because of the short

18

no-loss escape depth for photoelectrons in the range 10–40 eV, the UPS method is surface sensitive, sampling only several atomic layers.

For a metal plus an adsorbate, photoemission will include a joint contribution from both the metal and the adsorbate. Subtraction of the photoelectron energy distribution curves ⎡metal plus adsorbate minus that obtained for the clean metal⎤ yields a difference spectrum characteristic of photoemission from the adsorbate; in addition, characteristic intensity losses from the metal are seen as the density of states in the metal is modified by adsorption.

The power of UPS for studies of adsorptive bonding are well illustrated by the work of Demuth and Eastman[11] (Fig. 4), where difference spectra for C_2H_6, C_2H_4 and C_2H_2 adsorbed on Ni(111) are shown, in comparison with gas phase UPS spectra for the same molecules. In the lower panel of Fig. 4 a comparison of physically adsorbed C_2H_6 with $C_2H_6(g)$ is made, and the correspondence between the broad features due to σ_{CH} and σ_{CC} orbitals is good. In this comparison, the energy scales have been shifted to eliminate the effect of screening of final electron-hole states by the metal electrons. Thus, C_2H_6 in the physically adsorbed mode of bonding is very similar to $C_2H_6(g)$ in its orbital spectrum.

When the same type of comparison is made for chemisorbed C_2H_4 and C_2H_2, it is seen that the σ levels correspond well with the gas phase spectra, but the π-levels are shifted to higher binding energies by about 1 eV. On this basis, it is concluded that chemisorptive bonding of C_2H_4 and C_2H_2 occurs mainly via a π-interaction. In these cases, π-bonding to a single Ni atom is the preferred model, rather than the often postulated di-σ bonding to 2 - Ni atoms with destruction of the π-bond. The constancy in energy of

19

the various σ-levels upon adsorption is also consistent with < 20% re-hybridization of the sp^2 orbitals in C_2H_4 upon chemisorption by Ni(11ī), based on an SCF-LCAO calculation. π-bonding of this type is widely re-cognized in organometallic compounds; recently Zeise's salt (involving C_2H_4 π-bonding to one corner of a square planar Pt Cl_3 complex) has been studied by neutron diffraction[12] in order to locate the H's of the C_2H_4. The C_2H_4-hydrogens are pushed back by 0.18 Å from the plane of the C_2H_4 molecule in this compound, suggesting partial rehybridization of the carbon sp^2 orbitals. In addition, the C=C stretching frequency is decreased by ~6% implying that for distortion of this magnitude, the π-bond remains essentially intact in the ligand.

V. Kinetic Measurements of the Rate of Catalytic Reactions

In the last 10 years, much progress has been made in the measurement of the rate of catalytic reactions on surfaces. For dispersed catalysts, this advance has been due in part to the introduction of the practice of measuring specific catalytic activity, i.e., the rate of reaction per unit area of active catalyst. The active surface area may best be determined by a number of techniques involving chemisorptive uptake.[13,14] For catalysts being studied using ultrahigh vacuum methods such as Auger spectroscopy, ESCA, LEED, molecular beams, etc., the use of single crystals of high bulk and surface purity has been of major importance.

Boudart and coworkers have classified a number of catalytic reactions into two general categories - those which are sensitive to the catalyst surface structure (demanding reactions) and those which seem to be in-sensitive to catalyst surface structure (facile reactions)[15]. In the case of the hydrogenation of cyclopropane over Pt catalysts (a facile reaction) it has been found that the rate of this reaction per surface Pt atom is

20

essentially constant over a very wide range of catalyst particle size from highly dispersed Pt crystallites on Al_2O_3 or SiO_2 supports to Pt foil and Pt single crystals.[16] Results of this type help to establish a strong link between catalytic studies on dispersed and single crystal catalysts. It should be possible to employ many of the modern methods of surface physics and chemistry to the study of this and other similar reactions with assurance that the results will be applicable to the processes which occur on a practical catalyst.

A second example of the study of catalytic rates on single crystals comes from the work of McAllister and Hansen.[17] The decomposition of $NH_3(g)$ at high pressures has been studied on three single crystal planes of tungsten, starting with atomically clean tungsten in an ultrahigh vacuum environment. It was shown (Fig. 5) that the rate of the reaction follows the expression,

$$rate = A + B \, P_{NH_3}^{2/3}$$

The "A" process is thought to be the rate of N_2 desorption from the essentially fully covered W-N surface. The "B" process is thought to occur on a complete W-N adlayer with complex intermediates being produced which eventually lead to N_2 and H_2 products. It should be noted that the W(111) plane is more active than W(100) or W(110) for this decomposition process. No model is presently available to explain this structural sensitivity.

VI. Conclusion

In this talk, I have attempted to select examples of recent work which illustrate some of the methods employed for studying the character of catalytic sites, catalytic species, and chemical kinetics at catalytic surfaces. At present, we are just beginning to achieve new insights through the use of the newer methods of surface science for the study of catalysis.[18] The future is bright for the eventual understanding of the structural chemistry

involved in heterogeneous catalysis, and possibly for the application of
this knowledge to the design of better catalysts.

REFERENCES

(1) T. E. Madey, J. T. Yates, Jr., D. R. Sandstrom, and R.J.H. Voorhoeve, Treatise on Solid State Chemistry, Plenum Press, Vol. 6, Chapter 6 (1976), (Ed. N. B. Hannay)

(2) M. Boudart, A. Delbouille, E. G. Deroyane, V. Indorina, and A. B. Walters, J. Am. Chem. Soc., $\underline{94}$, 6622 (1972).

(3) R.J.H. Voorhoeve, J. Catal., $\underline{23}$, 236 (1971).

(4) L. H. Little, Infrared Spectra of Adsorbed Species, Academic Press, London (1966).

(5) M. L. Hair, Infrared Spectroscopy in Surface Chemistry, Marcel Dekker, New York (1967).

(6) R. P. Eischens, W. A. Pliskin and M.J.D. Low, J. Catal. $\underline{1}$, 180 (1962).

(7) R. J. Kokes, A. L. Dent, C. C. Chang, and L. T. Dixon, J. Am. Chem. Soc., $\underline{94}$, 4419 (1972).

(8) A. L. Dent, private communication.

(9) Y. Soma and W.M.H. Sachtler, Japan J. Appl. Phys. Suppl. 2, Pt. 2, 241 (1974).

(10) S. Hufner, G. K. Wertheim, J. H. Wernick, Phys. Rev. B8, 4511 (1973).

(11) D. E. Eastman and J. E. Demuth, Proc. 2nd Intern. Conference on Solid Surfaces, Japan J. Appl. Phys., Suppl. 2, Pt. 2, 827 (1974).

(12) W. C. Hamilton, Acta Cryst., 25A, 5172 (XIV-46) (1969).

(13) J. H. Sinfelt, Catal. Rev., $\underline{3}$, 175 (1970).

(14) J. E. Benson, H. S. Hwang, and M. Boudart, J. Catal., $\underline{30}$, 146 (1973).

(15) M. Boudart, A. Aldag, J. E. Benson, N. A. Dougharty, and C. G. Harkins, J. Catal., $\underline{6}$, 92 (1966).

(16) D. R. Kahn, E. E. Petersen, and G. A. Somorjai, J. Catal., <u>34</u>, 294 (1974).

(17) J. McAllister and R. S. Hansen, J. Chem. Phys., <u>59</u>, 414 (1973).

(18) J. T. Yates, Jr., Chem. and Engineering News, <u>52</u>, 19 (1974).

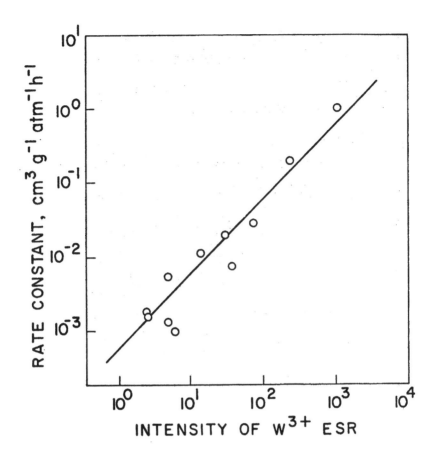

Fig. 1. Identification as W^{3+} as the active site in benzene hydrogenation over WS_2-based catalysts by the linear relation between the ESR intensity and the hydrogenation rate constant.

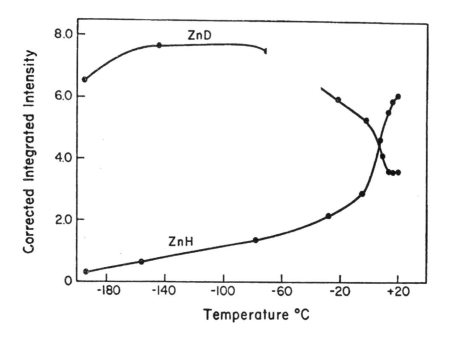

Fig. 2. Infrared measurement of the thermal equilibration of species
produced by adsorption of HD on ZnO at -195°C.

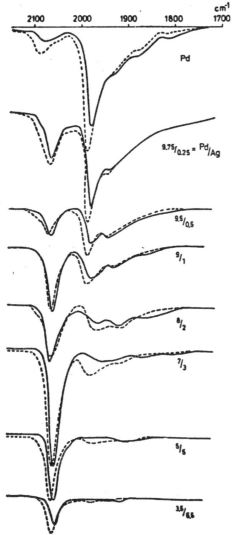

Fig. 3.   Infrared spectrum in the CO stretching region for CO adsorbed on Pd or Pd/Ag alloys. The IR band near 1950 cm^{-1} is assigned to bridge-bonded CO; the band near 2060 cm^{-1} is assigned to linear adsorbed CO.

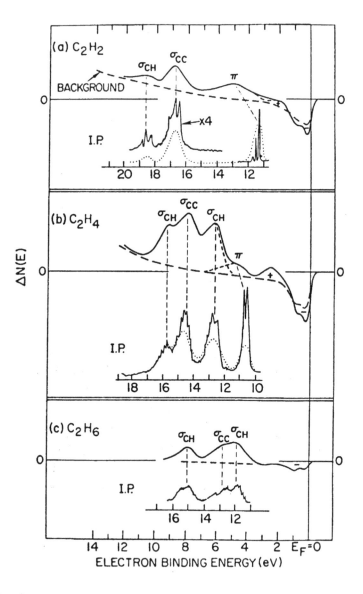

Fig. 4. Ultraviolet photoelectron difference spectra of organic molecules adsorbed on Ni(111).

28

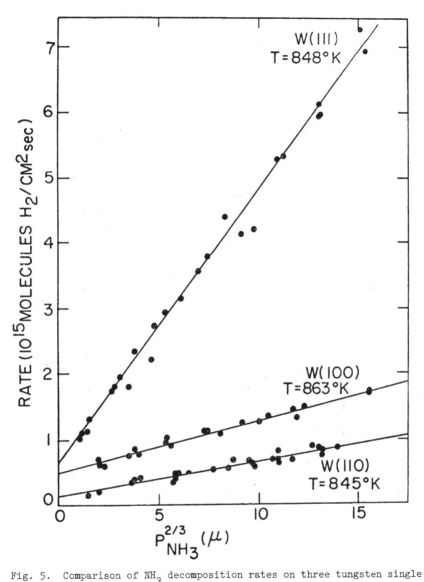

Fig. 5. Comparison of NH_3 decomposition rates on three tungsten single
crystals at similar temperatures. The NH_3 pressure is in units
of 10^{-3} Torr.

PANEL DISCUSSION: EXPERIMENTAL METHODS IN CATALYSIS

Chairman: Professor Gary Haller, Yale University

Recorder: Dr. Theodore E. Madey, National

Bureau of Standards

Panel Members:

Professor James Katzer, University of Delaware

Professor Robert L. Park, University of Maryland

Professor Thor Rhodin, Cornell University

Dr. John T. Yates, Jr., National Bureau of Standards

The session chairman, Professor Haller, began the discussion by introducing the panel members and asking each of them to define some of the problem areas in experimental methods related to catalysis. Following their opening remarks, Dr. Farrel Lytle presented a paper in which he discussed the utility of extended x-ray absorption fine structure (EXAFS) in studies of supported catalysts. Finally, the audience and panel participated in a free-wheeling dialogue concerned with modern methods and concepts in catalysis.

In the following pages, we have attempted to paraphrase the essence of these three distinct phases of the panel discussion. Except for the paper by Dr. Lytle, which is printed in the form he supplied, the remainder of the account is based on our (Haller, Madey, Yates) collective recollection.

I. Opening remarks by Panel Members

Professor Katzer noted that there is too frequently not much interaction between experiment and theory, and he called for stronger such interactions. He made a few remarks concerning the proper interaction between experiment and theory and suggested that the minimum interaction requires the use of fundamental chemical principles such as the laws of thermodynamics. Katzer pointed out that one of the major problems is that experimentalists measure what is most easily measured and theorists calculate what is most easily calculable, and frequently the two don't really get together. He proposed reducing catalysis to its most irreducible factor and suggested, following some earlier remarks of Professor John Turkevich, that d-electrons and transition metals are the most irreducible factor on the one side and protons on the other. He realized, of course, that this is an over-simplification. Another issue which he addressed concerned Professor Boudart's earlier discussion of the structure sensitivity of reactions, and he raised the issue as to what is the most irreducible state of a supported catalyst. Does it consist of 2 to 5 atoms or 5 to 10 atoms: just what is the smallest size of active catalyst? Characterization of the structure sensitivity of reactions necessarily involves larger samples involving clusters of many ($\sim$1000) atoms. These clearly present special problems for theorists because of the large number of atoms involved. He then turned from the discussion of solid state problems to surface problems, in the sense that he tried to define the most important step in catalysis. What is the rate limiting step in catalysis? Is it dissociation of the reactant (as in the case of ammonia synthesis) or is the rate controlling step a bimolecular reaction on the surface, or is the rate controlling step dissociation of stable surface intermediates? He noted that real progress and understanding of gas phase reactions came with the use of spectroscopic tools to follow the decomposition of intermediates. Similar studies are just now beginning on surfaces. He made a plea to bridge the gap between work on single crystals and real practical catalysts, between low and high pressures, as well as a

plea to bring experiment and theory closer together. He thought that
the theorists can best accomplish this by calculating trends as one goes
from molecule to molecule or sample to sample.

Professor Rhodin opened his comments by asking to what extent
does the quantum description of chemical bonding obtained from calculations
and electron spectroscopies of clean surfaces contribute to an understanding
of simple chemical reactions. The corollary question to this is: how
does the understanding of simple chemical reactions contribute to a better
understanding of industrial catalytic processes? He attempted to answer both
questions by stating that the study of simple reactions or adsorption on clean
surfaces allows one to develop and test new concepts, and that conceptual
principles can be applicable to practical catalysts. X-ray and vacuum ultraviolet photoemission, ion-neutralization, field emission, appearance potential,
Auger and electron loss spectroscopies all provide information on electron
structure of the solid surface and UV photoelectron spectroscopy provides
information on the chemical nature of adsorbed molecules as well. While all
spectroscopies are rather limited in theoretical understanding, the clarification of the chemisorption process on well defined metal surfaces is a first
step in understanding simple chemical reactions on metals. Rhodin cited
the work of Ertl on CO oxidation in high vacuum on clean metals as an example
of a simple chemical reaction which appears to attain the same characteristics
as when carried out under practical conditions ($\sim$ 10 orders of magnitude pressure
change). He indicated that this may be a situation of some generality for
reactions where the pressure ratio of reactants and products (and not absolute
pressure) is a critical factor. Rhodin believes it may be possible to make a
compilation of rate constants for elementary reaction steps on well defined
surfaces and these could be used in the analysis of a postulated sequence of

chemical reactions in a more complex set. He emphasized that measurement of the rate constants under dynamic conditions may be required. Rhodin concluded by restating that the study of the physics and chemistry of well defined systems can provide information on reaction rate mechanisms as well as the nature of chemical bonding of molecules at surfaces and these serve as building blocks in the development of new concepts. The role of critical design parameters in the engineering of new catalytic processes is essential but probably not directly amenable from a study of the physics and chemistry of well defined surfaces.

Professor Park pointed out that as a new technique appears on the horizon, there is a contrived enthusiasm for the technique which soon passes when people recognize that it can provide qualitative information but not quantitative information. There are few techniques which have proven to be "cure-alls" for the field. One example he used as an illustration is the difficulty in making Auger spectroscopy quantitative. Auger spectroscopy is fine for qualitative determination of surface cleanliness, but it is in-adequate for quantitative determination of surface composition. This only as an example - there are many examples where the experimental techniques we use simply are incapable of giving us quantitative information.

Professor Haller noted that the talk by Dr. Yates precluded the necessity of further opening remarks by him, and introduced Dr. Farrel Lytle.

Investigation of Supported Catalysts by
X-ray Absorption Spectroscopy[*]

by

Farrel W. Lytle
The Boeing Company
Seattle, Washington 98124

Summary of Remarks

The unique ability of x-ray absorption spectroscopy to isolate
one particular atom in a complex material, and from the energy position
and shape of the absorption edge determine the chemical state and from the
extended x-ray absorption fine structure (EXAFS) determine the radial arrangement of the atoms surrounding the absorbing atom, can be used to good
advantage in the study of heterogeneous catalysts. The absorption edge
spectroscopy has been summarized by Azaroff.[1] Theories of EXAFS which
more or less agree have been given by Stern,[2] Lee and Pendry,[3] and by
Ashley and Doniach.[4] The demonstration by Sayers, et al[5] that EXAFS can be
Fourier transformed into a radial distribution function surrounding the
absorbing atom has created a new interest in the technique for studying
glassy materials,[6,7] complex biological molecules,[8-11] solutions containing
coordination complexes,[12] gases,[13] and supported catalysts.[14,15]

The measurement and normalization of EXAFS using conventional
x-ray sources have been described[16] as have been the details of data processing.[17]
The recent advent of the EXAFS spectrometer[18] at the Stanford Synchrotron
Radiation Project (SSRP) now offers an unparalled x-ray flux (10^8-10^{10} photons
sec^{-1}) with a band width of 1 eV. The experiments described here were
performed on this instrument.[*]

The L_{III}-edge absorption[22] of Au, Pt, Ir, and Ta is shown in
Fig. 1. The data have been normalized to unit absorption and lined up on
the inflection point of the first rise on the low energy side of the edge.

34

These threshold resonance ("white lines") transitions are often observed[1]
in transition metal absorption spectra and are qualitatively understood
as allowed dipole transitions, 2p-to-empty 5d levels. Figure 1 shows this
in that the resonance is present in Au where all the 5d levels are filled
and increases with Pt, Ir and Ta as the 5d shell empties. For metals such
simple electron counts are unjustified as conduction bands are created.
The band structure calculations for Au, Pt, and Ir of Smith et. al[19] and
for Ta by Matthiess[20] were integrated in the region from the Fermi energy
to 10 eV above in order to more accurately estimate the unfilled d-density
of states and compare to the absorption resonance. This is summarized in
Table 1. In all cases the resonance (area or amplitude above the Au edge)
was found to increase with increasing d-states although the relationship
was not linear. The absorption spectra of the Au L_{III} edge (and the band
structure integration) was subtracted from that of each of the other elements
to isolate just the absorption to d-states. In Figure 2 data from Pt
compounds and 1 wt. pct. catalysts supported on Cabosil (SiO_2) is compared
to Pt metal. Data for $PtCl_2$ (not shown) was nearly identical in amplitude
and position with Pt metal. Again, the Pt compounds show resonance increases
as expected from an estimate of d-vacancies. The catalyst was sensitive to
its surface preparation. The "reduced" sample was prepared by a 500°C
reduction in flowing H_2, transferred to an air tight sample cell in a dry
box under N_2 and then measured. This same sample after exposure to air
comprised the other sample. The much smaller resonance in the reduced
sample may be due to filling of the Pt d-band by hydrogen as in the familiar
quenching of Ni magnetization by hydrogen.[2]

    In a similar 5% Pt on Cabosil sample the L_{III} EXAFS was measured
and a Fourier transform obtained. A typical example is shown in Figure 3.
The magnitude of the transform is plotted vs radial distance from the
absorbing Pt atom. It is seen that the environment of Pt in a well-dispersed
(90% by gas adsorption) catalyst is not simple. The identification of
various peaks with possible atomic species has been made considering the
state of the sample and the expected interatomic distances. Bonding to
oxygen and other Pt atoms is expected and found to vary as the sample was
oxidized or reduced. Cl was also expected in the sample. The "short bond"
may be evidence of epitaxy to the support. The distance $\sim 1.7\text{Å}$ is the same

35

as Si-O in the support and may be envisaged as Pt filling a missing Si site on the three-oxygen-atom "nest" of the (111) plane. This same kind of "short bond" has been found in supported Au catalysts.[15]

In summary the technique is a general and powerful one for the investigation of the electronic and structural environment of the catalytic atom. Planned experiments will use the high flux of the SSRP facility for in situ analysis of catalysts during reduction and various chemisorption experiments.

*For the experimental opportunity I·thank the SSRP staff and NSF and ERDA who fund the facility.

References

1. L. V. Azaroff, "X-ray Spectroscopy", McGraw Hill, N.Y. (1974).

2. E. A. Stern, Phys. Rev. B10, 3027 (1974).

3. P. A. Lee and J. B. Pendry, Phys. Rev. B11, 2795 (1975).

4. C. A. Ashley and S. Doniach, Phys. Rev. B11, 1279 (1975).

5. D. E. Sayers, E. A. Stern, and F. W. Lytle, Phys. Rev. Lett. 27, p.24, (1971).

6. D. E. Sayers, F. W. Lytle, and E. A. Stern in "Amorphous and Liquid Semiconductors", North Holland (1974), p. 403.

7. D. E. Sayers, E. A. Stern, and F. W. Lytle, Phys. Rev. Lett. 35, 584 (1975).

8. D. E. Sayers, F. W. Lytle, M. Weissbluth, and P. Pianetta, J. Chem. Phys. 62, 2514 (1975).

9. D. E. Sayers, E. A. Stern, and J. R. Herriott, J. Chem. Phys. 64, 427 (1975).

10. B. Kincaid, P. Eisenberger, K. O. Hodgson, and S. Doniach, Proc. Nat. Acad. Sci. USA 12, 2340 (1975).

11. R. G. Shulman, P. Eisenberger, W. E. Blumberg, and N. A. Stombaugh, Proc. Nat. Acad. Sci. USA (to be published).

12. P. Eisenberger and B. M. Kincaid, Chem. Phys. Lett. 36, 134 (1975).

13. B. M. Kincaid and P. Eisenberger, Phys. Rev. Lett. 34, 1361 (1975).

14. F. W. Lytle, D. E. Sayers, and E. B. Moore, Appl. Phys. Lett. 24, 45 (1974).

15. I. Bassi, F. W. Lytle, and G. Parravano, J. Catalysis (to be published).

16. F. W. Lytle, D. E. Sayers, and E. A. Stern, Phys. Rev. B11, 4825 (1975).

17. E. A. Stern, D. E. Sayers, and F. W. Lytle, Phys. Rev. B11, 4836 (1975).

18. B. M. Kincaid, P. Eisenberger, and D. E. Sayers, Phys. Rev. (to be published).

19. N. Smith, G. Wertheim, S. Hufner, and M. Traum, Phys. Rev. B10, 3197 (1974).

20. L. Matthiess, Phys. Rev. B1, 373 (1970).

21. R. Selwood, S. Adler, and T. Phillips, J. Am. Chem. Soc. 76, 2281 (1954): 77, 1462 (1955).

22. F. W. Lytle, J. Catalysis (to be published).

Table 1. Compilation of d-band information for pure metals, Pt compounds, and 1 wt. pct. Pt on Cabosil (silica) catalysts.

Material	Empty states, electrons/atom Isolated atom d-electron count	Band structure integration, 0-10eV		Area * of L_{III} threshold peak	Amplitude * of L_{III} threshold peak
		Total Unfilled States	Unfilled d-States		
Au	0	0.7	0.0	0	0
Pt	1	1.5	0.8	1.00	1.0
Ir	2	2.6	1.9	1.51	1.2
Ta	7	6.5	5.8	2.74	1.8
$PtCl_2$	2			1.10	1.1
$\alpha-PtO_2$	4			2.15	3.4
1% Pt, H_2 reduced $\sim$ (0.6)	}	estimated from results		0.69	0.6
1% Pt, exposed to air $\sim$ (1.0)		contained in this table		1.02	1.1

* Au absorption edge was subtracted from each of the L_{III} edges of the other materials.

38

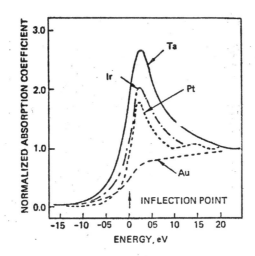

Figure 1. X-ray adsorption spectra of Au, Pt, Ir, and Ta near the L_{III} absorption threshold. The data were normalized by fitting a straight line 50–200 eV above the peak and extrapolating to below the edge to obtain the jump ratio. Point-by-point division of the L_{III} threshold absorption curves by the jump ratio produced these curves normalized to unit absorption. The zero of energy in each case is the 1st inflection point after the onset of absorption obtained by numerically differentiating the spectra.

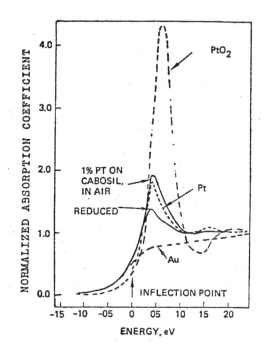

Figure 2.   X-ray absorption spectra of Pt catalyst samples and
            α-PtO$_2$ compared to Pt and Au.   Same normalization and
            energy scale as in Figure 1.

40

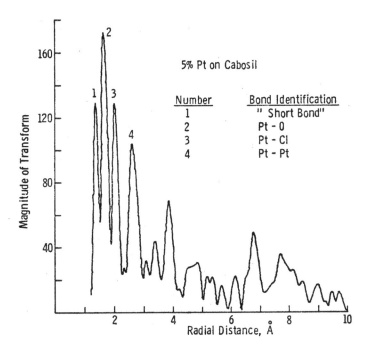

Figure 3. Fourier transform of Pt $_{LIII}$ EXAFS from a 5% Pt on
Cabosil sample which had been exposed to air.

III. Dialogue between Panel and Audience

Since Dr. Lytle's talk was fresh in the minds of the
audience, the first questions dealt with the utility of EXAFS in catalytic
studies. Dr. E. Siegel (Public Service and Gas and Electric) asked whether
or not one can study a real catalytic reaction using EXAFS. The conditions
he suggested are 800° Farenheit, 300 Atmosphere. Lytle answered that temperature is no problem but that pressures in excess of several atmospheres can
be a problem. The basic difficulty is that one needs an x-ray transparent
cell which is able to contain the sample, and he knows of no material strong
enough and transparent enough to be used for high pressure catalytic reactions.

Dr. Paul Citrin (Bell Telephone Laboratories) pointed out
that the EXAFS method seems to have a unique applicability in studies of
surfaces. That is, under certain circumstances, it may be made specifically
sensitive to surface species by observing the fine structure due to elements
in the adsorbed molecule rather than just to changes in substrate species.
He observed that one can detect not only the absorption edge due to a surface
species, but one can also monitor the photoelectrons ejected or the Auger
electrons ejected as one sweeps the wavelength of synchrotron radiation. He
suggests that specific surface EXAFS (SEXAFS) may provide unique information
about structural sensitivity of catalytic reactions.

Lytle noted that there will often be signal from molecules
not on the active sites of interest just as with other spectroscopies.   He
then observed that scientists from Bell Laboratories and U. California (Berkeley)
have also used the x-ray fluorescence being emitted from the atom absorbing
the x-rays as a very

sensitive detector in the EXAFS technique. He further pointed out that
Dr. Mel Kline (Berkeley) has looked at manganese in a leaf at concentrations
of the order of a few parts per million and has obtained an observable EXAFS
signal. Experimental times are of the order of one-half hour to 45 minutes per
absorption edge.

Dr. Warren Grobman (IBM) pointed out that one limitation
of EXAFS as applied to catalytic systems, is the fact that many of the atoms
of interest - carbon, oxygen, and hydrogen - are hard or impossible to
detect using this technique. The threshold for x-ray absorption is at a
fairly low energy and self-absorption by the substrate would frequently
cause attenuation of the signal due to surface species. This is particularly true for high Z substrates.

Citrin pointed out that the fluorescence technique and/or
the search for Auger electrons accompanying the x-ray absorption are the
ideal ways to study the adsorbed surface species.

Park said that for studying low Z materials the decay of
core hole excitations are such that the emission of Auger electrons is much
more probable than x-ray fluorescence.

Dr. Charles Duke (Xerox and the University of Rochester) issued
a warning concerning theoretical calculations. There are two types of calculations one can do. Firstly, one can compute ground state properties, reaction
probabilities, etc., and secondly, one can calculate excitation spectra. In
all of the spectroscopies discussed today, one is concerned with excitation
spectroscopy as a tool for studying ground state properties. Duke points
out that there may be little correlation between the excitation spectra and
the ground state properties and that model calculations appropriate for analyzing

one may be entirely inappropriate for the other. As an example, Hückel and CNDO/2 models are useful for the prediction of ground state properties (e.g., geometries, dipole moments) of organic molecules whereas spectroscopic CNDO/S models are required for interpretation of electronic spectra like optical absorption or photoemission. Thus, in quantum chemistry it is an accepted (even if undesirable) procedure to utilize different semi-empirical models to interpret different properties. Consequently, one should approach with caution the task of extracting ground state properties (e.g., geometries) from electronic spectroscopies (e.g., photoemission).

Duke also cautioned against simplistic assessments of the structure-determination capability of kinematic analyses of EXAFS data. To extract geometric information from kinematic analyses of such data, uncertainties in individual atomic electron scattering cross sections, the use of a finite data base, and multiple scattering phenomena must be assessed quantitatively. Such assessments have not yet been reported. Therefore, a substantial effort lies ahead before the present promise of EXAFS structure analysis is converted into a reality by the actual determination of previously unknown structures.

Lytle responded by noting that recent calculations by Lee and Pendry (Phys. Rev. B11, 2795 (1975) show that multiple scattering can be observed if one goes beyond the fourth coordination sphere but that multiple scattering is not a problem if you are only concerned with the first or second coordination spheres. The data truncation problems inherent in finite Fourier transforms have been handled using standard techniques.

Siegel concluded the discussion of EXAFS by noting that this technique may be particularly useful for studying high Z poisons on catalysts.

Seigel then asked questions concerning the role of
d-electrons in catalysis. Basically, he wanted to know: why are all
practical and important catalysts d-hole deficient?

Prof. Michel Boudart (Stanford University) answered by
pointing out that there are many catalysts which do not involve d-electrons
or d-holes. For example, protons, MgO and Alumina are all good catalysts
for certain catalytic reactions.

Seigel then asked: Could one envision the role of d-holes
as follows. In chemisorption, electron transfer must take place from the
adsorbate to the substrate and vice versa, and it is much easier when
electrons transfer from an adsorbed molecule to the substrate if there
is a d-hole rather than a filled d-orbital.

Prof. John Turkevich (Princeton University) suggested that
this idea may have some merit. In order to activate a molecule at the

surface, one simple-minded way of doing so is to take away electrons. This can be best accomplished if one can make a temporary transfer of electrons from the molecule to the solid substrate.

Dr. J. W. Gadzuk (NBS) suggested that one new promising technique for studying catalytic reactions may be the chemiluminescence technique pioneered by Kasemo in Sweden. He asked for comments from members of the panel. Park answered with a description of Kasemo's experiments. He pointed out that adsorption of oxygen on magnesium and aluminum gave rise to a yield of photons with a probability of 10^{-7} photons per adsorption event. He further speculated that chemisorptive luminescence might be a useful tool for following catalytic reactions even though the probability of a catalytic reaction might be a lot less than the usual high sticking probability for adsorption of molecules on surfaces. One might be able to integrate over a long time and follow the light emission from a sustained catalytic reaction. Gadzuk then pointed out that one advantage of chemisorptive luminescence as a tool for following catalytic reactions is the fact that the quantity detected is a photon, unlike other surface spectroscopies in which the quantities detected are charged particles. Thus, this technique appears to be applicable to high pressure catalytic situations. Park agreed.

Turkevich described some experiments in which they have generated hydrogen atoms by irradiation of molecular hydrogen dissolved in silica. The hydrogen atoms were detected using ESR spectroscopy. When the sample was heated to -150°C from LN_2, the hydrogen atoms recombined at impurity centers and light was emitted. There is a one-to-one correspondence between the disappearance of the ESR signal and the photons emitted. He suggests that studies involving light emission may be very useful in following energetic

46

catalytic reactions.

Prof. Ponec (Leiden) had several comments to make with respect
to John Yates' lecture. First of all, he observed that the hydrogenation of
cyclopropane is only structure insensitive on platinum, where the only reaction is the addition of hydrogen. He speculates that on nickel, where a
bond breaking reaction competes with hydrogenation, the reactions involving
hydrogen and cyclopropane are probably very structure sensitive. Secondly,
he took issue with the XPS studies of silver-palladium alloys. Although
the center of the bands did not change as the concentration of silver and
palladium in the alloys changed, there were significant changes in the band
edges. He pointed out that recent infrared studies by Sachtler indicate
that for CO on silver-palladium alloys, there are slight shifts in the
infrared absorption bands that had not been detected previously. That is,
slight shifts as a function of silver-palladium concentrations. In his
next comment, Ponec pointed out the danger of experimentalists trying to
follow theory too closely. He cited, as an example, the rigid band theory.
If one believed this model, then one should not be able to titrate copper
nickel alloy surfaces using hydrogen atoms but in fact there is preferential
absorption of the hydrogen on nickel in complete disagreement with the
concepts of the rigid band picture. He also indicated that Turkevich's
notion of activation of surface species by interaction of the absorbed
molecule with the d-holes or by electron transfer of the absorbed molecule
to the d-holes is an over-simplification of an incorrect idea by Dowden.

Gadzuk inquired as to whether or not it was feasible
or reasonable to study catalytic reactions using ultrahigh vacuum techniques
that is, to pump the system to low pressures following a catalytic reaction

and examine the state of the surface. To what extent does this sort of procedure help in the understanding of the actual catalytic mechanism?

Yates replied that first of all, one should study the kinetics of the reaction while the gas is in the system and not simply rely on the measurement of the surface following the reaction. Secondly, he pointed out that catalytic intermediates may be removed as the gas is pumped away from the sample. On the other hand, one can then examine residues, such as carbon, following the catalytic reaction. If carbon is present on the surface at that stage, then it is logical to assume that the carbon was also present during the course of the reaction.

Dr. Galen Fisher (NBS) inquired: what specific experimental techniques have been useful for determination of catalytic intermediates?

Katzer answered that infrared absorption techniques are probably the most useful and most widely applied method for looking at catalytic intermediates. However, it is often difficult to distinguish between actual reaction intermediates and other stable adsorbed species. Katzer also issued a warning concerning the use of surface spectroscopies. He pointed out that these techniques are useful for studying chemisorption processes as well as for looking at the most stable intermediates. However, they are generally not useful for studying the dynamics of catalytic processes. Fundamental information concerning reaction mechanisms can probably best be obtained using such techniques as modulated molecular beam methods.

Haller pointed out that infrared and other spectroscopies can be useful for studying catalytic intermediates, provided that you combine such techniques with the kinetic transient method. The objective here is to correlate the change in concentration of surface species with changes that are occuring in the gas phase in a transient experiment. In this way, one can be reasonably certain that you are studying something which is kinetically important

Park indicated that both optical reflectance and chemisorptive luminescence would be useful for studying catalytic intermediates under actual high pressure and transient conditions. He further suggested the use of isotopic labelling in kinetic studies as a means of getting information about intermediates. (He did not mention the work of Emmett using radioactive ^{14}C tracers, but that is an interesting instance of the applicability of this technique to the detection of intermediates).

Fisher said that in many cases, there may be multiple pathways to products with very stable reaction intermediates which exist under certain conditions but which may not be the most kinetically important intermediates.

Finally, Yates suggested a restatement of the question: can one detect the transition state in a catalytic reaction? He indicated that in homogeneous kinetics, it frequently occurs that the concentration of the transition state may be very, very low, but one may hope to deduce the structure of the transition state if the structure of intermediate precursors can be determined.

Boudart commented that many of the old conventional chemisorption methods and methods of chemical analysis can frequently be used to calibrate some new spectroscopies such as Auger spectroscopy. In particular, he mentioned an example from his own laboratory in which they used quantitative chemisorption of CO, CO_2 and N_2 to calibrate some Auger spectroscopic measurements on ammonia synthesis catalysts. Prof. Boudart further pointed out that such techniques as EXAFS should best be done on well-characterized samples, on samples which have been studied in several different laboratories. He mentioned also that there is a program now, scientific interchange in matters of catalysis between the US and the USSR, and that this program is providing a mechanism for exchange of catalyst samples that have been characterized in different laboratories.

Prof. Theodore Einstein (Univ. of Md.) asked: how conclusive is the evidence for the adsorption of atoms in high symmetry sites as is usually assumed in low energy electron diffraction calculations? He based his skepticism on the recent electron stimulated desorption ion angular distribution (ESDIAD) measurements made by Madey, Czyzewski and Yates at the National Bureau of Standards and on the calculations of Gersten, et al. which indicated that some ion desorption patterns can be explained on the basis of adsorption at sites not associated with high symmetry positions on the substrate.

Park indicated that it has not been established theoretically that adsorbed atoms and molecules usually sit in sites of high symmetry. The LEED calculations which have been performed to date suggest that high symmetry adsorption sites are appropriate in some cases. However, because of the type of model calculations usually performed by LEED theorists, it is impossible to test all possible adsorption sites. Perhaps wider use of inversion techniques will demonstrate this more clearly.

Haller then called for closing remarks by the panelists.

Yates reformulated the question that he had asked at the end of his talk. If one has knowledge of the atomistic details of catalytic reactions such as specific adsorption sites, knowledge of surface structures and intermediates, knowledge of specific reaction rates under carefully controlled conditions, how can this information be transferred in a practical sense to assist in the design and construction of a catalytic reactor for a specific practical catalytic experiment? He acknowledged that the instrumental methods of the surface scientists are being more widely used by catalytic chemists (e.g., ESCA is finding increasing use in catalytic studies). The real question concerns the utility of basic concepts as applied to practical catalyst design.

Katzer emphasized the need to understand the chemical composition of complex catalytic systems, and observed that modern methods provide this information.

Park suggested that "educated intuition" plays a major role in catalyst design, and basic research on surfaces provides catalytic chemists with better models on which to base their intuition.

Finally, Rhodin cautioned that kinetics are essential in catalysis, and modern methods should be used in conjunction with kinetic measurements.

Dr. Lawrence Bennett (NBS) closed the panel discussion by thanking the participants and noting that there were some techniques that were not considered in any detail. One technique is the Mossbauer effect, which Professor Boudart has shown has usefulness in real catalytic situations, and in which one can hope to distinguish between the surface and bulk particles. Another technique which was not mentioned is one which is being developed at NBS - perturbed angular correlation. It is a specialized technique which measures the same type of thing that Mossbauer techniques do, namely, the hyperfine fields, but it is also restricted to elements with specific nuclear properties. In the case of iron, the NBS group has shown that information can be obtained by

combining perturbed angular correlation and Mossbauer experiments that are
not observable from either method alone because of resolution problems.
Another isotope which is very useful in perturbed angular correlation is
rhodium, and that may be of some interest in catalysis. There was not any
discussion about nuclear magnetic resonance, but he thinks that is another
technique which has a great deal of usefulness in catalysis.

EFFECTS OF
ALLOYING

Moderator:
H.EHRENREICH
Harvard University

Lecture by:
J.SINFELT
Exxon

Session 2.

Effects of Alloying

Panel Members:

C. D. Gelatt, Jr.,
Harvard Univ.

V. Ponec,
Gorlaeus Lab., Leiden

J. Sinfelt,
Exxon

Recorder:

A. J. McAlister,
NBS

SESSION 2. EFFECTS OF ALLOYING

Chairman: H. Ehrenreich, Harvard University

Recorder: A. J. McAlister, N.B.S.

Panel Members: J. H. Sinfelt, Exxon
C. D. Gelatt, Harvard University
V. Ponec, Gorlaeus Lab.

The session on the Effects of Alloying was formally divided into two parts: an invited talk by Dr. Sinfelt on the topic "Catalysis by Alloys and Bimetallic Clusters"; and a panel discussion, chaired by Dr. Ehrenreich, during which Dr. Ponec and Dr. Gelatt made short formal presentations, and in which the audience participated actively.

CATALYSIS BY ALLOYS AND BIMETALLIC CLUSTERS

(A brief summary of Dr. Sinfelt's remarks, with
selected figures and general references).

I. Surface Enrichment

It has long been realized that the surface composition of an alloy
may differ from that of the bulk, but only in recent years have attempts
been made to obtain information about the composition of the actual
surface exposed to reacting molecules. Chemisorptive titration and
Auger spectroscopy have been the principle experimental tools employed
in such studies, and at least qualitative agreement has been obtained
with theory, which predicts, in rough terms, that the component with
lower heat of vaporization in the pure state will be enriched on the
alloy surface.[1]

II. Specificity of Metals and Alloys

Catalytic activity depends strongly on the reaction considered.
Two reactions were used as examples: a) ethane hydrogenolysis, in which
carbon-carbon bond rupture is believed rate determining, and for which
adjoining active metal atoms are believed necessary as surface sites: b)
cyclohexane dehydrogenation, for which desorption of the product benzene
is believed to be rate determining.

A strong correlation has been observed between rates of ethane
hydrogenolysis over Group VIII and IB metals and the Pauling percent d-
character in the bonds of the metals. This correlation, shown in Figure
1, is perhaps the strongest available indication of the importance of an
electron factor in catalysis by metals.

The variation in the rates of the example reactions over Cu-Ni
alloys of about 0.1% dispersion is shown in Figure 2. The decrease in

56

hydrogenolysis activity as Cu is added to Ni probably involves a strong structure factor; if for instance two adjacent metal atoms are needed, these become scarcer as Cu is added. This structure factor is enhanced by the surface enrichment of Cu indicated by hydrogen chemisorption measurements on the samples. Cyclohexane dehydrogenation activity first increases with Cu addition, suggesting a decrease in the strength of product binding. It then remains roughly constant, decreasing only at high Cu content, with the rate determining step perhaps moving back through the reaction sequence to the initial chemisorption of cyclohexane.

The data above were obtained on alloys of metals, which are miscible, at least at high temperatures. It is of interest to consider what happens if the metals are immiscible in bulk, for example Cu and Ru. One might expect the alloy in this case to consist of mixed particles of pure Cu and Ru, and in view of the results of Figure 1, to find no change in specific activity as Cu is added to Ru. In fact, Cu has a marked affect on surface processes occurring on Ru. As Figure 3 shows, for unsupported Cu-Ru catalysts, of about 1% dispersion, both the volume of strongly chemisorbed hydrogen (i.e., that fraction which cannot be pumped off at room temperature) and the rate of ethane hydrogenolysis are markedly reduced by the addition of Cu. Hence the Cu appears to go onto the Ru surface much as if it were chemisorbed there.

III. Dispersion Effects in Bimetallic Systems

Suppose we prepare highly dispersed, supported bimetallic samples, for example by co-impregnation and subsequent heat treatment. Will we obtain mixtures of pure metal particles, or, as for the unsupported low dispersion samples cited above, find interactions between the different metals whose effects are manifested in chemisorption and catalytic activity? Figures 4 and 5 show the effects of adding Cu to Ru and Os on H and CO uptake and on the rate of ethane hydrogenolysis. A strong interaction between the Group VIII and Group IB metals is indicated.

57

Now, how do the results on bimetallic systems for low and high dispersion tie together? In Figure 6 are plotted the data for hydrogen uptake, and in Figure 7, for hydrogenolysis activity, as a function of composition, for large Ru-Cu aggregates (dispersion ∿1%) and for highly dispersed Ru-Cu clusters (dispersion ∿50%). These results are consistent with the notion that Cu covers the Ru particles; when the dispersion is greater, a given fraction of Cu will cover a much smaller percentage of the particle surface.

IV. Hydrogenolysis Activity and Strong Chemisorption

Figure 8 shows a striking correlation between ethane hydrogenolysis activity of Ru-Cu catalysts and their capacity for strong hydrogen chemisorption. This suggests that we lie on the left side of a volcano curve, where there is a positive effect between strength of binding and catalytic activity.

V. Selectivity of Bimetallic Catalysts

We finally consider the selectivity of bimetallic catalysts. In particular, we look at the rates of cyclohexane dehydrogenation into benzene, and hydrogenolysis into low molecular weight fragments, principally methane, over Ru-Cu catalysts. We see in Figure 9 that the addition of Cu to Ru enchances the selectivity to the production of benzene by an order of magnitude.

REFERENCES

[1] See for example, R. A. Von Santen and M. A. M. Boersma, J. Catal. 34, 13 (1974) and references therein.

[2] J. H. Sinfelt, Cat. Rev. - Sci. Eng. 9, 147 (1974) and references therein.

[3] J. H. Sinfelt, Y. A. Lam, J. A. Cusumano, and A. E. Barnett, J. Catal. 42, 227 (1976).

[4] J. H. Sinfelt, J. Catal. 29, 308 (1973).

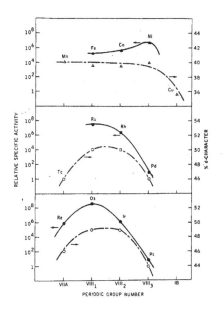

Figure 1. Catalytic activities of metals for ethane hydrogenolysis in
relation to the percentage d-character of the metallic bond. The closed
points represent activities compared at a temperature of 205° C and
ethane and hydrogen pressures of 0.030 and 0.20 atm, respectively, and
the open points represent percentage d-character. Three separate fields
are shown in the figure to distinguish the metals in the different long
periods of the Periodic Table. (From Ref. [2])

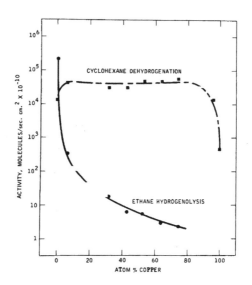

Figure 2. Activities of copper–nickel alloy catalysts for the hydrogenolysis
of ethane to methane and the dehydrogenation of cyclohexane to benzene. The
activities refer to reaction rates at 316° C. Ethane hydrogenolysis activities
were obtained at ethane and hydrogen pressures of 0.030 and 0.20 atm,
respectively. Cyclohexane dehydrogenation activities were obtained at
cyclohexane and hydrogen pressures of 0.17 and 0.83 atm, respectively.
(From Ref. [2])

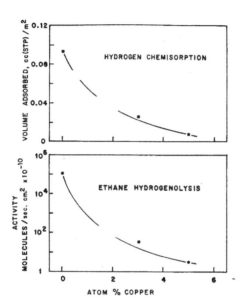

Figure 3. Hydrogen chemisorption capacity and ethane hydrogenolysis activity of ruthenium-copper catalysts as a function of copper content. The hydrogen chemisorption data were obtained at room temperature and represent the strongly chemisorbed fraction. The ethane hydrogenolysis activities are reaction rates at 245° C and ethane and hydrogen pressures of 0.030 and 0.20 atm, respectively. The catalysts were prepared by heating in hydrogen at 500° C. (From Ref. [3]).

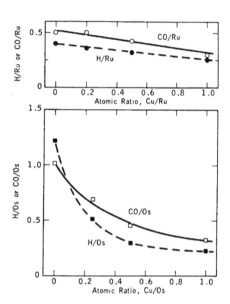

Figure 4.   The chemisorption of hydrogen and carbon monoxide at room
temperature on silica-supported ruthenium-copper and osmium-copper catalysts.
The catalysts all contain 1 wt% ruthenium or osmium, with varying amounts of
copper.   The adsorption data are expressed by the quantities H/Ru, CO/Ru, H/Os,
and CO/Os, which represent the number of hydrogen atoms or carbon monoxide
molecules chemisorbed per atom of ruthenium or osmium in the catalyst.   (From
Ref. [2])

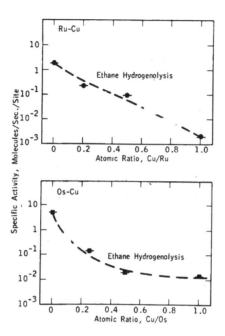

Figure 5. The specific activities of supported ruthenium-copper and osmium-copper catalysts for ethane hydrogenolysis. Activities are shown for the same catalysts used in obtaining the chemisorption data in Figure 4. The activities are compared at 245° C and ethane and hydrogen pressures of 0.030 and 0.20 atm, respectively. Specific activity is defined here as the activity per surface site, and is determined by dividing activity per atom of ruthenium or osmium in the catalyst by the quantity H/Ru or H/Os, respectively, from Figure 4. (From Ref. [2])

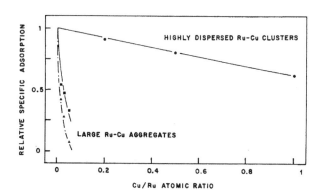

Figure 6. Influence of the state of dispersion of ruthenium-copper catalysts
on the relationship between hydrogen chemisorption capacity and catalyst
composition. The square and triangular points represent total hydrogen
chemisorption and strongly chemisorbed hydrogen, respectively, on the
large ruthenium-copper aggregates. (From Ref. [3])

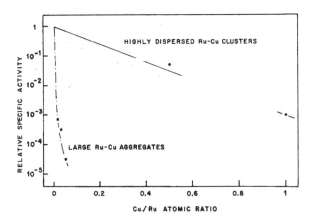

Figure 7.   Influence of the state of dispersion of ruthenium-copper
catalysts on the relationship between ethane hydrogenolysis activity
and catalyst composition.   The large ruthenium-copper aggregates
have a metal dispersion of the order of 1%, while the highly dispersed
ruthenium-copper clusters have a metal dispersion of the order of
50% (From Ref. [3])

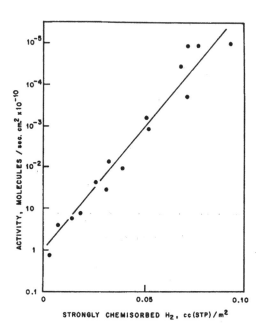

Figure 8. Correlation of ethane hydrogenolysis activity and amount of strongly chemisorbed hydrogen for ruthenium-copper catalysts. The ethane hydrogenolysis activities are rates at 245° C and ethane and hydrogen pressures of 0.030 and 0.20 atm, respectively. (From Ref. [3])

66

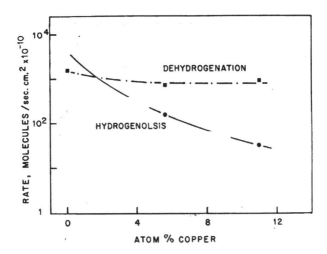

Figure 9. The rates of dehydrogenation and hydrogenolysis of
cyclohexane on ruthenium-copper catalysts as a function of composition.
The rates are shown for a temperature of 316° C and for cyclohexane
and hydrogen pressures of 0.17 and 0.83 atm, respectively. The catalysts
were reduced at 400° C. (From Ref. [3])

ALLOYING PANEL DISCUSSION

The chairman opened the discussion with a request for comments and questions on Dr. Sinfelt's talk. Dr. Ponec remarked with regard to hydrogenolysis on Cu-Ni alloys: when Cu is added to Ni, a multiple Cu-Ni bond can still be formed, the main effect of alloying being on the size of ensembles of Ni sites; for cracking reactions (e.g. hexane), Ni and Cu give terminal cracking, while Pt gives mid-cracking; interestingly, Ni-Cu alloys give Pt-like mid-cracking. Dr. Ponec then asked Dr. Sinfelt if he could give any insight as to why d-character might be important in catalysis. Dr. Sinfelt replied that he simply regarded this correlation as an observed fact for hydrogenolysis, and that it is of predictive value for such reactions. He suggested the possibility that absorbate-metal bonds might resemble those of the bulk metal in some respect. Dr. Gelatt noted that correlation of % d-character with hydrogenolysis activity is no stronger than that of such a geometric factor as the volume per atom of the transition metals, the latter correlating as well with hydrogenation data as does % d-character. Dr. L. H. Bennett noted that surface enrichment or otherwise depends on preparation, and asked Dr. Sinfelt if reactions could be studied as a function of surface composition. Also to this point, Prof. Rhodin asked whether the experimental environment might affect surfaces; could the catalytic surface be different during reaction from that prepared? Dr. Sinfelt felt that surfaces can indeed change during the catalytic process, but probably did not change much in the systems he discussed in his talk. Prof. Ponec cited some H_2 and CO titrations that showed chemisorption to affect surface composition of Cu-Ni catalysts; he

68

suggested that here Ni was drawn to the surface. Prof. Turkevich showed H_2 adsorption data on 32Å particles of Au-Pt alloys in which the quantity absorbed varied nearly linearly from a maximum at Pt to nothing at Au. Prof. Turkevich noted that this was at variance with the cherry model, (see below, particulary Prof. Ponec's contribution) since bulk phase diagrams show distinct phase change, and concluded that phase relations must differ in small particles.

Prof. Park raised the question of whether the local character of the electronic structure of Cu and Ni in Cu-Ni alloys were not in fact rather like that in the host metals. Dr. Gelatt replied that in fact, this was largely the case according to detailed calculations within the coherent potential model. (For example, see Fig. 4 of Dr. Ponec's contribution which shows the local density of states at a Ni site in a CuNi alloy.)

Dr. B. J. Evans noted that about 10Å resolution was available in electron microscopy, and wondered why one could not obtain direct evidence of surface enrichment in highly dispersed systems. Dr. Sinfelt and Dr. Ponec replied that this was not adequate resolution for the problem.

Dr. Lagally asked what experimental evidence there might be of persistence of elemental local electronic structure character in alloys. Prof. Ertl briefly described some soft x-ray appearance potential spectra from his laboratory which confirm this picture. Dr. Siedle asked what was known of the morphology of the bimetallic clusters. Dr. Sinfelt replied that at present, one only knows that they are very small and composed of both metals, and really nothing more. He hoped that the extended x-ray absorption fine structure technique (see the previous session) would contribute much to the solution of this problem.

Dr. Ponec then presented a short review of the experimental situation regarding the phase composition and surface enrichment of Cu-Ni alloys. His notes on this presentation are reproduced below.

SURFACE COMPOSITION OF NICKEL-COPPER ALLOYS
(CONTRIBUTED DISCUSSION BY DR. V. PONEC)

The knowledge of the phase composition is essential for the discussion to follow. Let me, therefore, mention several facts in this respect first.

As can be seen from Fig. 1, several phase diagrams have been suggested in the literature. Because of the evident uncertainty of the information on the phase composition of Ni-Cu alloys at catalytically interesting temperatures, 150-400° C, Franken[6] from our laboratory reinvestigated this problem with evaporated metal films and x-ray diffraction. He found that at 420° C the films consisted of only one phase after a rather short time; at 215° C one-phase films were formed when the equilibration was performed during a sufficiently long period (about 40 hours); and at 165° C two phases persisted even after still longer sintering. Now, we know also from some other additional experiments that the critical temperature lies between 165-200° C, as predicted by Meijering[7].

Once converted into a one-phase system, the films do not show any detectable segregation of phases at low temperatures (20-100° C) even after several days. However, it is known[8] that segregation can be induced at 20° C by an electrochemical formation and vacuum decomposition of Ni-Cu hydrides. After such segregation, the Cu-rich β-phase forms the surface of the whole system[8]. So much for the phase composition.

Sachtler and Dorgelo[9] and Van der Plank and Sachtler[3] suggested the use of work function (φ) measurements and selective chemisorption of

71

H_2 to determine the surface composition of Ni-Cu alloys. The main result of these measurements is: films equilibrated at 200°C revealed a broad range of bulk concentration for which the surface concentration was approximately constant (indicated by both mentioned methods) and this "constant" composition was approximately equal to the bulk composition of β-alloys (as indicated by the H_2 chemisorption). The explanation of the authors[1-3] was: two phases coexist at this temperature and the β-phase forms a shell around the α-phase crystallites ("cherry" model).

However, doubts and criticism on this model appeared soon. First, the coexistence of two phases at 200° C was not certain[6,7]. Further, according to the rigid band theory (RBT) and Dowden's ideas[10], Ni in alloys with 60% and more Cu cannot adsorb hydrogen; the chemisorptive titration was actually "theoretically impossible". On the other hand, the work function measurements which are undoubtedly sensitive for the surface composition are not supported by any theory which would provide us with a theoretical relation for ϕ as a function of composition. Many expectations were, therefore, related to the Auger spectrometry.

Three groups of authors investigated Ni-Cu alloys by Auger spectrometry and they all reported the same final result: alloys equilibrated at 400-500° C (one-phase alloys) reveal the same bulk and surface composition; no indication of a constant surface-composition with varying bulk-composition has been obtained[11-13]. However, in contrast to these data, selective hydrogen chemisorption (see Fig. 2) suggested a surface composition constant in a wide range of bulk composition and, moreover, the same composition for films equilibrated at 200° C and powders equilibrated at 400° C[14-16]. This has led to reconsideration

72

of all data and analysis of the assumptions made. We came to the following conclusions[6,17]. (1) In spite of the use of ϕ-measurements for
the surface determination being empirical, one conclusion is apparently
always right: where ϕ is constant, the surface composition is constant.
In this light the following result of Franken[6] was important - see
Fig. 3. The materials equilibrated at 420° C[6], analogous to powders
used by other authors[14-16], reveal the same behaviour as that observed
by hydrogen chemisorption (compare Figs. 2 and 3). Further, the one-
phase (equilibrated at 215° C, 420° C) and two-phase (165° C) films
showed the same function ϕ. The new data by Franken[6] reproduced
the old data by Sachtler and Dorgelo[9] very closely. (2) The theoretical objections against the chemisorptive titration by hydrogen have
been essentially removed by the new coherent potential approximation
theory[18,19]. We can see that this theory predicts that the local Ni
density of states in the alloy is similar to that in the pure metal.
Further, a comparison of the data in Fig. 4 with the data for pure Ni
shows that when for chemisorption the presence of certain states is
necessary, these states are always found on Ni atoms, never on Cu atoms,
also at highest dilution. This gives us confidence in chemisorptive
titration of surface Ni atoms. (3) When suspicions against the two
mentioned methods were eliminated, attention had to be concentrated on
Auger spectrometry. Two groups[22,23] reinvestigated the problem and
they showed that alloys equilibrated at 400-500° C do not reveal a
substantial surface enrichment when Auger electrons of E $\sim$ 800 eV are
used for analysis. However, when Auger electrons of E < 100 eV are used
a clear enrichment in Cu is detected, approximately up to the values
derived from hydrogen chemisorption! In this way, the previously existing

73

controversy has been removed and the surface composition of Ni-Cu alloys is now reasonably well established. Recently, Brongersma has confirmed these results by an independent method – low energy ion scattering[27].

Experience has thus taught us how valuable measurements of ϕ can be for information on surface composition. It is to be regretted that so far no theory exists relating the variation of the surface dipole layer (inferred from variation in ϕ) to the bulk and surface composition. There is hope that the new methods[18-21, 24-26] applied in the theory of surface states will provide such information if the theoreticians accept the challenge.

REFERENCES

[1] Sachtler, W. M. H. and Jongepier, R. J., J. Catal. 4 (1965) 665.

[2] Vecher, A. A. and Gerasimov, J. I., Russ. J. Phys. Chem. 37 (1963) 254.

[3] Van der Plank, P. and Sachtler, W. M. H., J. Catal. 12 (1968) 35.

[4] Rapp, R. A. and Maak, F., Acta Met. 10 (1962) 62.

[5] Elford, L., Müller, F. and Kubaschewski, O., Ber. Bunsenges. 73 (1969) 61.

[6] Franken, P. E. C., Thesis, Leiden University, 1975. Franken, P. E. C. and Ponec, V., J. Catal., in press.

[7] Meijering, J. L., Acta Met. 8 (1957) 257.

[8] Palczewska, W. and Majchrzak, S., Bull. Acad. Polon. Sci. Série Chim. 17 (1969) 681.

[9] Sachtler, W. M. H. and Dorgelo, G. J. H., J. Catal. 4 (1965) 654.

[10] Dowden, D. A., J. Chem. Soc. 1950, 242; Ind. Eng. Chem. 44 (1952) 997.

[11] Harris, L. A., J. Appl. Phys. 39 (1968) 1419.

[12] Quinto, D. T., Sundaram, V. S. and Robertson, W. D., Surface Sci. 28 (1971) 504.

[13] Ertl, G. and Küppers, J., J. Vac. Sci. Technol. 9 (1972) 829.

[14] Cadenhead, D. A. and Wagner, N. J., J. Phys. Chem. 72 (1968) 2775.

[15] Ponec, V. and Sachtler, W. M. H., J. Catal. 24 (1972) 250.

[16] Sinfelt, J. H., Carter, J. L. and Yates, D. J. C., J. Catal. 25 (1972) 283.

[17] Ponec, V., Catal. Rev. - Sci. Eng. 11 (1975) 41.

[18] Soven, P., Phys. Rev. 156 (1967) 809; 178 (1969) 1136.

[19] Velický, B., Kirkpatrick, S. and Ehrenreich, H., Phys. Rev. 175 (1968) 747.

[20] Hüfner, S., Wertheim, G. K. and Wernick, J. H., Phys. Rev. $\underline{B8}$ (1973) 4511.

[21] Stocks, G. M., Williams, R. W. and Faulkner, J. S., Phys. Rev. $\underline{B4}$ (1971) 4390.

[22] Helms, C. R., J. Catal. $\underline{36}$ (1975) 114.

[23] Yamashina, T., Watanabe, K., Fukuda, Y. and Hasiuba, M., Surface Sci. $\underline{50}$ (1975) 591.

[24] Kalkstein, D. and Soven, P., Surface Sci. $\underline{26}$ (1971) 85.

[25] Haydock, R., Heine, V., Kelly, M. J. and Pendry, J. B., Phys. Rev. Letters $\underline{29}$ (1972) 13.

[26] Berk, N. F., Surface Sci. $\underline{48}$ (1975) 289.

[27] Brongersma, H. H., Private Communication (Philips).

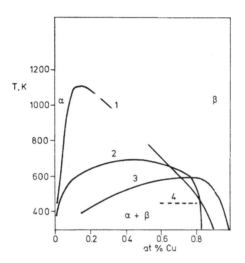

Fig. 1. Phase diagrams for Cu-Ni suggested by various authors.

1 - Refs. [1] and [3], calculations according to data in Ref. [2].

2 - Ref. [3], calculations according to data in Ref. [4].

3 - Ref. [5].

4 - Refs. [6] and [7].

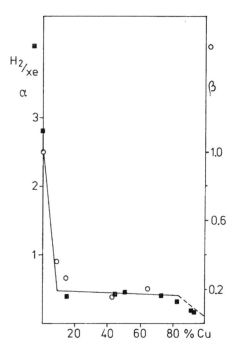

Fig. 2. Selective chemisorption of hydrogen by Ni-Cu alloys.

α=hydrogen chemisorption/xenon physisorption ratios for films equilibrated at 200° C, according to Ref. [15]. (squares)

β=hydrogen adsorption/cm^2 of alloy, Ref. [16]; β = 1 for Ni. (circles)

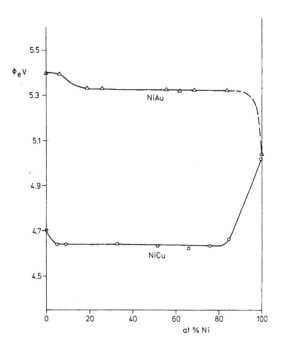

Fig. 3. Work function measurements[6] on Ni-Cu films equilibrated at 420° C
(the films equilibrated at 165° C and 215° C show exactly the same
values).

For comparison: data for a similar Ni-Au system (two-phase systems
at all temperatures used here).

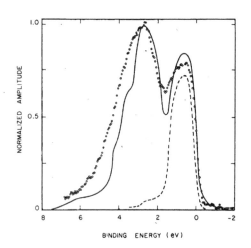

Fig. 4. Comparison[20] of the experimental XPS valence band structure (dots) with the CPA theory (full line) for an alloy of 40% Ni in Cu. Schematically, the density of states localized at Ni atoms, as calculated by CPA[21] is indicated by the dashed curve.

80

CONTINUATION OF THE PANEL DISCUSSION

Dr. Cohen asked whether it was not possible to construct (a) non-equilibrium surfaces and (b) do catalytic measurements on them. Prof. Boudart replied that such surfaces can be formed. He cited work by Helms, also by Yowashita, in which surfaces were prepared by high energy ion sputtering and annealing, and Auger measurements used to show their non-equilibrium surface compositions. Dr. Russo inquired if, although there is no theory for interpretation of work function changes, there are not at least some semi-empirical rules established for this task. Dr. Ponec said there was, citing published studies from his laboratory.

A query was made about possible substrate effects on metal clusters. How do support effects occur? Are there any electronic effects of the support on clusters. Dr. Sinfelt replied that in general, such effects should be possible, although for the systems he discussed, supported and unsupported samples gave about the same results (when it was possible to study them in this way). But he felt it would not be hard to visualize cases where such effects could occur.

Dr. Breiter pointed out that in his experiments on bulk, solid Au-Pt alloys, H_2 adsorption occurs only on Pt rich phases.

In reply to Dr. Sinfelt's last remarks, it was noted from the floor that Moss found that bimetallic clusters of Pd and Ag could be formed on silica, but not on alumina supports, implying some strong interaction with the alumina.

Dr. Yates asked Dr. Sinfelt if some carbon residue builds up in the course of his rate measurements; do characterization tests give the same results before and after reactions. Dr. Sinfelt replied that

81

surely some residue occurs, but it should not be large. His practice is to monitor activity continuously, and the rates do not drift with time.

Prof. Ehrenreich commented on the reasons for a lack of theoretical analysis of work function change measurements: (a) electronic structure calculations for alloys are hard; (b) surface calculations are harder; (c) dipole layers involve many body effects which are harder still.

Next, Dr. Gelatt presented the results of a systematic calculation of the band structure and heats of formation of bulk, stoichiometric transition metal hydrides across the 3d and 4d rows. The motivation for presenting these bulk results was two fold: in the preparation of supported transition metal catalysts, there is always a high temperature, H_2 reduction step, under conditions suitable to the formation of hydrides; secondly, it is of interest to try to understand transition metal hydrogen bonding in a simpler system than surface adsorption. This work is presented in the following reprint from Solid State Communications, 17, 663 (1975). (A more detailed description of these calculations is in preparation.)

HEATS OF FORMATION OF 3d AND 4d TRANSITION METAL HYDRIDES*

C.D. Gelatt, Jr., Jacquelyn A. Weiss and H. Ehrenreich

Division of Engineering and Applied Physics, Harvard University, Cambridge, MA 02138, U.S.A.

(Received 3 March 1975; in revised form 23 April 1975 by E. Burstein)

One electron energy spectra are used to explain the heats of formation of stoichiometric transition metal hydrides across the 3d and 4d rows. The trends agree reasonably with existing experimental information. The magnitudes are predominantly (but not exclusively) determined by the formation of a metal—H bonding band. In contrast to the screened proton model, the result is not directly related to the Fermi level density of states.

THE IMPORTANT factors contributing to the heat of formation of transition metal hydrides have been identified on the basis of band theory and used to obtain a systematic understanding of the chemical trends across both the 3d and 4d rows. Three of the principal ingredients are: (1) the formation of a metal—hydrogen bonding band; (2) the lowering of the metal d bands; (3) the binding of the additional electron associated with each hydrogen atom at the top of the Fermi distribution. In contrast to the screened proton model,[1] systematic trends across the transition metal rows are not determined by the variation of the Fermi level density of states.

Results for the band structure of PdH_x for various concentrations are shown in Fig. 1. Switendick[2] has previously studied the energy bands of stoichiometric hydrides without, however, calculating the heats of formation or the energy spectrum of disordered non-stoichiometric hydrides. The present results are based on the Korringa—Kohn—Rostoker (KKR)[3] approach and renormalized atom potentials.[4] The complex energy bands for non-stoichiometric hydrides were obtained using the average t-matrix approximation (ATA),[5] as extended to a rocksalt structure (for example, $\underline{CuH}$[6]) with randomly distributed hydrogen atoms and vac-

ancies. The results for Pd[7] and stoichiometric PdH[2] are in good agreement with those previously published.

In the dilute hydrides [Fig. 1(b)] a Pd—H molecular bonding level appears below the Pd band structure [Fig. 1(a)]. Figure 1(b) shows that some of the bands are damped due to the disorder as indicated by the width of the shading. Within the ATA levels below the muffin-tin zero, such as the PdH bonding level, are not damped. With increasing hydrogen concentration the molecular level broadens into a band, while simultaneously the spectral density of the lowest Pd band becomes increasingly broad and weak. In the stoichiometric hydride [Fig. 1(c)] this band has been replaced by one associated with the Pd—H molecular level in the dilute case. Structure corresponding to this band has been observed in β-phase PdH photoemission experiments.[8]

The heat of formation per unit cell ΔE for the reaction of metal M with hydrogen gas to form hydride MH_x is

$$M(\text{solid}) + \tfrac{1}{2}xH_2(\text{gas}) \rightarrow MH_x(\text{solid}) - E. \quad (1)$$

The change in energy is therefore

$$\Delta E = E(MH_x) - E(M) - \tfrac{1}{2}xE(H_1), \quad (2)$$

where $E(MH_x)$ and $E(M)$ are respectively the total energies per unit cell of the hydride and the pure metal. $E(H_2) (= -2.266 \text{ Ry}$[9] in the Hartree—Fock approxi-

* Research supported by the National Science Foundation under grants GH-32774 and DMR 7203020.

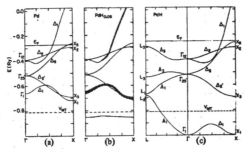

FIG. 1. Energy bands for PdH$_x$ along the [100] direction for $x = 0, 0.05$, and 1. Also shown for $x = 1$ are the bands along [111].

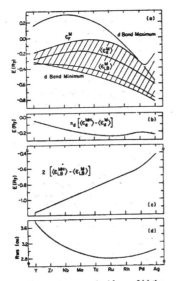

FIG. 2. The variation across the 4d row of (a) the upper and lower d band edges, the average energy of the occupied d bands, $\langle \epsilon_d^M \rangle$, and the metal Fermi energy, ϵ_F^M; (b) the shift in average energy of the occupied d bands; (c) the occupation weighted shift in average energy of the lowest band; and (d) the Wigner–Seitz radius, R_{WS}.

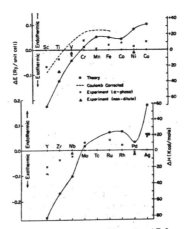

FIG. 3. The calculated heats of formation, ΔE, for stoichiometric hydrides without (solid line) and with (dashed line) Coulomb energy corrections. The additional points represent experimental enthalpies of formation, ΔH, of α-phase (reference 12) and non-dilute (reference 13) 3d and 4d hydrides.

mation) is the energy required to separate a hydrogen molecule into its constituent protons and electrons.

We shall approximate ΔE by the difference in the sum of one-electron valence band energies $\Delta E_1 \equiv$

$E(MH_x) - E(M)$. The three principal contributions to ΔE_1 for the stoichiometric case are given by the empirical formula

$$\Delta E_1 = 2(\langle \epsilon_{LB}^{MH} \rangle - \langle \epsilon_{LB}^{M} \rangle) + n_d(\langle \epsilon_d^{MH} \rangle - \langle \epsilon_d^{M} \rangle) + \epsilon_F^{M}. \quad (3)$$

They consist of (1) the shift in average energy of the lowest band (LB) containing two electrons, (2) the average shift of the d bands multiplied by n_d, the number of d electrons contained in the occupied d bands above the lowest band, and (3) the energy at which the added electron in the unit cell enters the solid, approximated by the Fermi energy of the metal.[10] Equation (3) can be used completely empirically, or the terms can be evaluated by the use of Brillouin zone integrations utilizing, for example, the special points averaging technique.[11] The results presented here are based on a two special-point average.

The variation of each of the terms of Equation (3) for f.c.c. metals and rocksalt structure hydrides across the 4d row is illustrated in Fig. 2. The behaviour of the d band width, shown in Fig. 2(a), is associated with the variation of the Wigner–Seitz radius, R_{WS} [Fig. 2(d)]. As R_{WS} decreases on the left side of the period, $\langle \epsilon_d^M \rangle$ rises and the bands broaden.[4] This behaviour, together with the filling of the d-bands, accounts for the initial rise in ϵ_F. On the right side of the period, the increase of R_{WS} and nuclear charge Z causes the d bands to shift downward and to narrow. These effects compete with d band filling to produce a net lowering of ϵ_F with increasing Z. The sharp rise of ϵ_F near Ag is a result of filling conduction band states above the top of the d bands.

The shift in d band energies, shown in Fig. 2(b), correlates with the amount of metal charge lying in the interstitial region. The decrease on the left side of the row is associated with filling of bonding d orbitals and increasing d band width: greater delocalization accompanies larger bonding. On the right side the effects of d band narrowing and concomitant d wave function localization dominate. The decrease near Ag is associated with the conversion of conduction band states in the metal to d band states in the hydride. It is apparent from Fig. 2(c) that the nearly linear shift in the average energy of the lowest band is the dominant and most rapidly varying contribution to the heat of formation.

The heats of formation of both 3d and 4d stoichiometric monohydrides calculated from equation (3)[10]

are shown in Fig. 3. In the absence of any systematic studies of monohydrides, we show experimental points for dilute hydrogen solutions[12] and selected non-dilute hydrides.[13] As a result, comparison between theory and experiment should be limited to an examination of general trends (or Z dependence) across the period. The gross features of the ΔE curve reflect the general behavior of hydride formation, namely that with the exception of Pd and Ni stable concentrated hydrides form only on the left side of the period. It is clear from Fig. 2 that the shape of the curve is determined largely by the variation of the Fermi energy and the lower band shift with Z. In particular, the dip near Pd and Co is attributable to the first of these effects.

The discrepancy on the left side of the periods is predominantly due to the neglect of the Coulomb repulsion associated with the increased charge density near a proton. This can be estimated by suitably defining an effective hydrogen volume. The prescription used here defines the hydrogen sphere radius to be located where the sum of metal and hydrogen atomic wave-functions has zero slope along the [100] direction. The number of electrons contained within this sphere is $n_H = 1.23, 1.15. 1.08$. and 1.00 for Ti, Cr, Fe, and Ni respectively. The Coulomb energy is $\frac{1}{2} n_H (n_H - 1) F_0$ when $n_H > 1$ and zero otherwise. F_0 is the standard Slater Coulomb integral. The results of including this estimate in the 3d row are shown by the dashed curve in Fig. 3. The effect is to bring the calculations into somewhat better accord with the heats of formation suggested by experiment for elements at the left end of the period. The correction vanishes to the right of Co. Experience with YH indicates that similar results can be expected for the 4d row. We interpret the large Coulomb energies of TiH (and YH) as indicating that the formation of a polyhydride (for which the estimated Coulomb energies are substantially smaller) may be more favorable, as is observed experimentally. Switendick's[2] calculations for CaF_2 structure dihydrides indicate that a second metal–hydrogen bonding level is formed, implying that the change in one-electron energy is compatible with $E(MH_2) - E(M) - E(H_2) < 0$.

The present calculations indicate no significant change in the charge density within the metal muffin-tin sphere upon hydride formation. To the extent that the metal charge density is properly monitored by this estimate, the effects of charge transfer appear to be small. However, even small charge transfer can influence the heat of formation significantly[5,14] and affect the trends. We also note that the neglect of metal Coulomb energy changes on hydride formation is supported by the preceding argument if the redistribution of charge in the metal sphere is sufficiently small.

The lattice strain energy, computed from the volume expansion and the bulk modulus, is ~ 0.01 Ry. Although this quantity is too small to affect ΔE appreciably, it is important in determining the phase diagram of the various metal hydrides.[15]

Despite these reservations, we should note that the predictions of the present calculations are considerably superior to those of the screened proton model[1] and do not involve adjustable parameters. The binding energies of hydrides are weak on the scale of $\frac{1}{2}E(H_2) =$ 1.13 Ry [equation (2)] and the discrepancies between theory and experiment on this scale are actually fairly small.

Acknowledgements — The authors would like to thank Dr. K. Krıtayakırana for assistance in the early stages of this work, and Prof. G.W. Pratt, Jr., and Drs. A. Bansil and H.L. Davis for help with the NaCl structure programs.

REFERENCES

1. EBISUZAKI Y. & O'KEEFFE M., *Prog. Solid State Chem.* 4, 187 (1967).

2. SWITENDICK A.C., *Solid State Commun.* 8, 1463 (1970); SWITENDICK A.C., *Int. J. Quantum Chem.* 5, 459 (1971); SWITENDICK A.C., *Ber. Bunsenges. Phys. Chem.* 76, 535 (1972).

3. KOHN W. & ROSTOKER N., *Phys. Rev.* 94, 1111 (1954).

4. HODGES L., WATSON R.E. & EHRENREICH H., *Phys. Rev.* B5, 3953 (1972); GELATT C.D., Jr., Unpublished Ph.D. thesis, Harvard University (1974); GELATT C.D., Jr., EHRENREICH H. & WATSON R.E. (to be published).

5. BANSIL A., EHRENREICH H., SCHWARTZ L. & WATSON R.E., *Phys. Rev.* B9, 445 (1974).

6. HUISMAN L.& WEISS J.A., *Solid State Commun.* 16, 983 (1975).

7. MUELLER F.M., FREEMAN A.J., DIMMOCK J.O. & FURDYNA A.M., *Phys. Rev.* B1, 4617 (1970).

8. EASTMAN D.E., CASHION J.K. & SWITENDICK A.C., *Phys. Rev. Lett.* 27, 35 (1971).

9. PARR R.G., *Quantum Theory of Molecular Electronic Structure*, Benjamin, NY (1972).

10. For the noble metals it is necessary to modify equation (3) by substituting for ϵ_F^M the explicit energy ($\sim \epsilon_F^{MH}$) at which the electron is added.

11. BALDERESCHI A., *Phys. Rev.* B7, 5212 (1973); CHADI D.J. & COHEN M., *Phys. Rev.* B8, 5747 (1973).

12. MCLELLAN R.B. & OATES W.A., *Acta Met.* 21, 181 (1973).

13. LIBOWITZ G.C., *The Solid State Chemistry of Binary Metal Hydrides*, Benjamin, NY (1965); MEULLER W.A., BLACKLEDGE J.P. & LIBOWITZ G.C., *Metal Hydrides*, Academic, NY (1968).

14. GELATT C.D., Jr. & EHRENREICH H., *Phys. Rev.* B10, 398 (1974).

15. ALEFELD G., *Ber. Bunsenges. Phys. Chem.* 76, 746 (1972); WAGNER H. & HORNER H., *Adv. Phys.* 23, 587 (1974).

CONCLUSION OF THE PANEL DISCUSSION

With reference to Fig. 3 of the reprint, Dr. Gelatt observed that bulk bonding of hydrogen is weak in the region of active catalysts, as one might expect.

Dr. Ponec questioned the validity of this analogy. He pointed out that Mo has one of the highest heats of adsorption for hydrogen, but that it does not form a hydride. He suggested that some factor related to the rupture of metallic bonding must be added to the present considerations.

It was enquired from the audience whether the potential for the calculations included core-core interactions, or was it all orbital. Dr. Gelatt replied that it was all orbital; the systems are metallic, and there is little evidence of charge transfer.

Dr. Gadzuk wondered about the apparent lack of importance of many body effects to the bulk calculations, in contrast to the frequency with which many body effects are invoked in chemisorption calculations. Dr. Gelatt replied that it was not clear that a good one-electron job has yet been done on the chemisorption problem.

Dr. Siegel noted that the calculated d band widths appear to remain the same upon hydride formation. Dr. Gelatt pointed out that in the NaCl structure assumed for the calculations, the states which define the top and bottom of the d bands are of a symmetry which does not see the proton.

Dr. Messmer noted that Yang, Johnson, and he have done cluster calculations, with hydrogen at tetrahedral and octahedral sites, and note the same d band width effect. But he noted that binding energies are different for hydrogen at the separate sites.

Dr. Duke asked Dr. Messmer if his cluster geometries were self consistently calculated or fixed. Dr. Messmer said they were fixed at the bulk geometry.

At this point, the chairman asked for comments as to whether these calculations are relevant to the problems of interest, or useful in some way; they are certainly easier to make than calculations of the real thing. Dr. Ponec made one such remark already. Were there others?

Dr. Bennett pointed out that bulk hydrogen seems to affect the catalytic properties of Raney Ni.

Dr. Sinfelt remarked that calculational insight into the ways molecules are adsorbed would be quite useful.

Prof. Boudart noted the importance of knowing about the bulk electronic structure of a variety of interstitial compounds, hydrides, carbides, and nitrides-some of which are active catalysts - with a view to predictions of new, useful catalytic materials.

Dr. Duke wondered if, since a high temperature reduction in hydrogen is always employed in the preparation of small supported particles, the hydrogen might not be incorporated in some essential way in their structure. This possibility was acknowledged by Dr. Sinfelt, and again the tenuous nature of our knowledge of the structure of these small particles was emphasized.

GEOMETRICAL EFFECTS

Moderator:

R.HANSEN

Ames Laboratory, Energy

Research & Development Admin.

Lecture by:

G.ERTL

University of Munchen

Session 3.

Geometrical Effects

Panel Members:

G. Ertl,
Univ. of Munchen

C. B. Duke,
Xerox Corp.

J. Turkevich,
Princeton

Recorder:

A. J. Melmed,
NBS

WORKSHOP ON THE ELECTRON FACTOR IN CATALYSIS

SESSION 3: GEOMETRICAL EFFECTS

THE RECORDER'S STORY

Allan J. Melmed

National Bureau of Standards
Washington, D.C.   20234

In a large sense, discussion of "geometrical effects' took place at
many random times throughout this Workshop as well as during the approximately two hours which were formally devoted to the subject. If one
attended the Workshop with the expectation of learning about geometrical
effects on the electron factor in catalysis, however, one perforce would
be disappointed. For, despite the wealth of discussion of geometrical
effects, there was a dire paucity of comments on geometrical effects in
catalysis and an even less densely populated set of comments attempting
to relate geometry of surfaces to something one might recognize as an
"electron factor."

It is appropriate to ask whether it is possible to separately
investigate electron and geometry effects for real surfaces. In order
to achieve a separation, it would be necessary to use surfaces with
constant geometry and varying electron factors and surfaces with
constant electron factor and varying geometry. In a strict sense, this
does not appear to be possible. The closest approach to these conditions
seems to be the use of surfaces of materials with similar crystal structure
in order to examine electron factor effects and the use of various
surface modifications of one material in order to examine geometrical
effects. But neither of these approaches actually separates the electron
factor and the geometrical factor. The hope is that in the two types of
experimentation, one factor or the other will be the dominant factor.
Probably a more accurate description would be to speak of the material
factor.

As a Recorder, it was not my responsibility to react emotionally to the
Workshop discussions. For those attendees who might have suffered some
disappointment, however, I will first attempt to explicitly recall those

91

isolated comments which somehow directly related geometrical effects to an _electron factor_ (Part 1). Then will follow a much larger collection of material which related to geometrical effects in surface physics, surface chemistry, and surface metallurgy; the formal contribution of the invited speaker, G. Ertl (Part 2), and the contributions from the panel (Parts 3-6). Finally, general discussion items are recorded (Part 7) and a succinct summary is given (Part 8).

Part 1.   Discussion Related to Geometrical Effects
on an Electron Factor

V. Ponec mentioned E. W. Müller's explanation of certain aspects of image contrast in the field ion microscope in terms of directed, or dangling bonds protruding from the surface (Z. Knor and E. W. Müller, Surf. Sci. 10 (1968) 21). (In the model referred to, the probability of electron transfer from an image gas molecule or atom to the specimen surface is strongly influenced by the geometry of the unpaired, virtual bonds of the surface atoms, and also by the degree of occupancy of these orbitals.)

G. Ertl, during his formal lecture (which follows), discussed Smoluchowski's ideas relating electron work function differences to differences in surface atomic geometry. He then discussed experimental results which might be partially understood in terms of such electron work function differences.

Part 2. The Geometric Factor in Chemisorption and
Catalysis on Metals

G. Ertl

Institut für Physikalische Chemie,
Universität München

I. Introduction

The idea that the geometric arrangement of the atoms in the surface
of a catalyst might be of importance for its activity and selectivity
emerged already in the early days of systematic research on the catalytic
action. The role of this principle becomes strikingly evident in the
field of biology, in which reactions of complex molecules are catalyzed
with extreme specificity by enzymes leading to the picture that these
catalysts fit their substrates, just as a key fits into a lock.

H. S. Taylor[1] suggested that similar effects might be of considerable
importance for reactions catalyzed by solid surfaces. He introduced the
concept of "active centres" which were believed to be regions characterized
by particular configurations of the surface atoms. Their actual geometry
was discussed particularly by Balandin[2] in the framework of his so-
called "multiplet theory."

Systematic experiments on these effects with "real" catalysts are
complicated by the fact that they consist of polycrystalline material
exposing different crystal planes and a whole spectrum of structural
imperfections, i.e., without any well-defined surface topography. It is
only possible to investigate this phenomenon by studying the catalytic
activity with particles of varying mean size[3]. Very small particles
(with diameters less than about 50Å) are predicted to expose a larger
number of atoms exhibiting a high degree of unsaturated valences. In
particular special adsorption sites at which an adsorbed particle is
surrounded by five surface atoms (B5 sites) were predicted to occur
predominantly with particles of about 25Å diameter[4]. As a result of
experimental studies in this direction surprisingly many catalytic
reactions were found to be structure-insensitive ("facile" reactions)[3],
and even in those cases in which the rate and selectivity were dependent

94

on the particle size ("demanding" reactions) it is not always clear
whether these effects are really due to the operation of the geometric
factor[3]. Since most of these investigations were performed with fcc
transition metals, at least for this class of materials, the conclusion
may be drawn that their catalytic activity with most reactions is not
influenced dramatically by the surface morphology.

Systematic studies can be performed by using single crystal surfaces,
probably themselves exhibiting well-defined structural imperfections
(arrays of steps), and apparently some of the recent results from this
latter approach seem to indicate a considerable influence of the surface
geometry--in contrast to most of the investigations on the particle-size
effect.

The rate of a catalyzed reaction, r, is in the simplest way determined
by the rate constant $k(T)$ and by a function of the concentrations of the
reacting species $f(c_i)$, i.e.,

$$r = k(T) \times f(C_i).$$

If the reacting particles are in equilibrium with the fluid phase
through adsorption--desorption steps (which is frequently the case),
their surface concentrations depend exponentially on the adsorption
energies E_i; therefore $f(c_i) = g(e^{E_i/RT})$. The rate constant may be
written as

$$k(T) = k_o \cdot e^{\Delta S|/R} \cdot e^{-E*/RT}$$

in which $\Delta S|$ denotes the activation entropy and $E*$ the (apparent) activation
energy. The latter will, in general, also be related to the adsorption
energies of the reacting particles. In the simplest case this will be a
linear relationship similar to the Brønsted law in acid-base catalysis.
As a consequence, even relatively small variations of the adsorption
energies (all the other parameters being constant) are predicted to
appreciably influence the reaction rate due to their exponential relationship.
From a simplified point of view, it may, therefore, be predicted that the
geometric factor will be of importance in the following cases:

   a) if relatively strong variations of the energy of the bond
       between the surface and the adsorbate are observed;

95

b) with reactions of larger molecules or with cooperative processes, i.e., if not only a single surface-adsorbate bond is involved in the elementary step of the reaction. Enzyme catalysis is an extreme example for this situation which is expected essentially to affect also the activation entropy. In the following, this aspect will, however, not be discussed in more detail.

According to a rather naive picture, one would assume that the strength of the adsorbate-substrate bond increases as the number of "unsaturated" valences of the surface atoms increases, i.e., if their coordination number decreases. This assumption generally does not hold as will be shown by several examples. (The energy for dissociation of an H-atom from H-O-H is 119 kcal/mole, but only 102 kcal/mole from H-O![5].) Using the picture of a surface molecule, variations of the adsorption energy are to be expected if the surface geometry influences markedly either the energies and/or occupancies of those orbitals of the surface atoms which are involved in the bond, or the overlap between the adatom orbitals and the corresponding group orbitals from the surface atoms. It can be assumed that these effects are more pronounced with solids exhibiting strongly directed bonds in the bulk which probably also persist at the surface ("dangling bonds"). The occurrence of electronic surface states represents a further complication of this very rude picture.

Bonds of this latter type are, for example, present with the elemental semiconductors (Ge, Si) which crystallize with the diamond lattice, and in fact a reaction with clean germanium surfaces has been found to be highly structure-sensitive[6]: Rather heterogeneous surfaces were created by crushing thin Ge slabs in UHV with a magnetically operated hammer. The decomposition of N_2O leading to gaseous N_2 and oxygen remaining attached to the surface served as a test reaction. Since oxygen does not desorb, the O atoms formed continuously block the "active sites" where the reaction takes place. As a consequence during the progress of the reaction the activation energy increases continuously from 7 to 45 kcal/mole (although even in the latter case the oxygen coverage was still far below saturation) thus demonstrating very strong variations of the catalytic activity across the surface (Fig. 1).

96

II. Metal Single Crystal Surfaces

With the transition metals variations of the electronic properties with the surface orientation become evident from theoretical as well as from experimental investigations.

Using a tight-binding approximation Hydock and Kelly[7] calculated the local densities of d-states at atoms in different low index surfaces. Figure 2 shows the results for the (110), (100), and (111) surfaces of a bcc crystal which have to be compared with those reproduced in Fig. 3 for the three most densely packed surfaces of an fcc lattice.

Ultraviolet photoelectron spectra from a bcc metal (W)[8], as well as from an fcc metal (Ni)[9], exhibit pronounced differences between differently oriented surfaces. Moreover, the W(100) surface is the first example of clear evidence for the existence of metallic surface states[10]. However, at least in the case of nickel, the anisotropy of the chemical behaviour is much smaller than would probably be expected on the basis of the different electron energy distributions.

The adsorption energies for hydrogen on different W single crystal planes, as measured as a function of coverage by Domke, et al.,[11] are reproduced in Fig. 4. Similar data are reported by Schmidt[12]. The initial heat of adsorption varies between 32 and 40 kcal/mole. Since the adsorption is dissociative this means that the strength of the metal-hydrogen bond varies between 68 and 72 kcal/mole, i.e., only by about 6%. At higher coverages the adsorption energies change due to energetic heterogeneities or due to the onset of repulsive interactions.

Much stronger differences were reported for the adsorption kinetics of nitrogen on tungsten. Adams and Germer[13] concluded that the sticking coefficient at room temperature is appreciably high only on those planes which contain sites with fourfold coordination for the adsorbate. In particular for the W(111) face (which is suspected to play a dominant role in ammonia synthesis at iron catalysts) the sticking coefficient is reported to be quite small[14].

One of the few examples reported in the literature, for which the rate of a catalytic reaction has been studied with different single crystal surfaces under UHV conditions is the work of McAllister and

97

Hansen[15] on the ammonia decomposition on tungsten. Some of their results are reproduced in Fig. 5 and indicate that the rate of NH_3 decomposition on W(111) is considerably higher than on the (100) and (110) surfaces. This appears to be in some contrast to the findings on the adsorption kinetics of N_2 on tungsten, since on the (111) plane the sticking coefficient is rather low[14], although a direct comparison between the kinetic data of these two different processes is problematic. Since the kinetic laws describing the rate of ammonia decomposition are rather complicated, it is not possible to see in a simple manner in which elementary step of the reaction the plane specificity comes into play. It is interesting to notice that the (apparent) activation energies on the (111) and (100) planes are found to be nearly identical but on the (110) surface are appreciably higher.

The most important catalytic reaction occurring at a bcc metal is certainly the synthesis of ammonia over iron catalysts. Despite enormous efforts in the past[16], the mechanism of this reaction is still unclear, although mostly the adsorption of nitrogen is considered as being the rate-determining step. Some years ago Brill, et al.,[17] suggested from some rather qualitative observations by means of the field emission microscope, that the NH_3 formation takes place preferentially on the (111) faces. This assumption was recently supported by some work from Boudart's laboratory[18] using small iron particles. These authors concluded--mainly based on the magnetic properties and CO adsorption data of their small catalyst particles--that the so-called C_7-sites as present on the Fe(111) surface are the most active ones in ammonia synthesis.

This picture is confirmed by very recent studies on the kinetics of nitrogen adsorption on clean Fe(100)[19] and (111)[20] single crystal surfaces using Auger electron spectroscopy to monitor the surface concentration. The variation of this quantity with the N_2 exposure is reproduced in Fig. 6 for both surface orientations. Whereas on the (100) surface the initial sticking coefficient is rather small (in the order of 10^{-7}), on the (111) surface chemisorption proceeds more rapidly by at least one order of magnitude. By means of low energy electron diffraction the formation of ordered adsorbate structures could be observed, but no nitrogen induced surface reconstruction as has been suggested to occur

with small catalyst particles[18] or field emission tips[17] was observed.
We thus believe that ammonia synthesis on Fe catalysts represents, in
fact, an example of a structure-sensitive reaction. However, whether
the C_7-sites play indeed the dominant role still remains somewhat specula-
tive.

Variations with the surface orientation of the initial adsorption
energies of hydrogen on nickel[21], as well as of CO on nickel[22] and
palladium[23] are listed in Tables 1-3 and may serve as examples for the
role of the geometric factor in chemisorption on fcc metals. With the
H/Ni system the differences are particularly small and the strength of
the metal-H bond is quite similar to that in the diatomic NiH molecule[24].
A similar behaviour is found for the adsorption of CO on nickel for
which again the adsorption energies are comparable to the dissociative
energy (35 kcal/mole[25]) of $Ni(CO)_4$. It is felt that this close correspondence indicates that cluster calculations (using the SCF-Xα technique)
are a successful approach to a theoretical treatment of chemisorption on
metals.

The last column of Table 2 contains the maximum number of CO molecules
adsorbed per cm^2 at room temperature and with CO pressures below 10^{-4}
Torr. Although the density of surface atoms varies between the three
planes by about 60% the adsorbed amounts are quite similar, irrespective
of any features characterizing the coordination of numbers of "dangling"
bonds of the surface atoms. Moreover the maximum densities of adsorbed
CO molecules with all fcc metal surfaces which were studied so far, are
determined by the tendency for the formation of close-packed layers,
whereby, an effective diameter of about 3A has to be attributed to the
adsorbed CO. This aspect is of some importance for those techniques in
which the metallic surface area of small catalyst particles is derived
from selective CO adsorption.

A series of LEED observations on the adsorption of CO on fcc metals[22,23,26]
revealed that saturation of the adsorbate layer is achieved by a continuous
compression of the unit cell of the adsorbate. This means that fixed
adsorption sites do not exist but rather that the binding energy changes
only slightly along certain directions on the surface.

Such a behaviour is also predicted on the basis of energy profiles
calculated by means of a modified Anderson-Grimley formalism under the

assumption of maximum overlap between the $2\pi^*$-orbital of CO and the metallic d-orbitals[27]. As an example the theoretical energy profile for CO on Pd(110) is reproduced in Fig. 7 together with the structural models (corresponding to increasing coverage) as derived from LEED observations. Aside from being a satisfactory explanation of the experimentally derived surface configurations of the CO molecules, this semi-empirical theory also predicts only relatively small variations of the binding energies with surface orientation which is also in agreement with the experimental findings.

The catalytic oxidation of CO over different Pd single crystal surfaces, as well as with a polycrystalline wire, was studied in some detail in our laboratory[28,29]. The conclusions on the kinetics of this reaction were recently confirmed by a series of papers by White and coworkers[30]. Figure 8 shows the variation of the steady-state rate of CO_2 formation with temperature at constant partial pressures of the reactants with different Pd surfaces. Obviously, there is no noticeable influence of the surface crystallography on the reaction rate. At temperatures below about 200 °C the rate is determined by the desorption of CO. (CO_{ad} inhibits the dissociative adsorption of O_2 which is a necessary prerequisite for the reaction to occur). From Table 3 it becomes evident that the heat of CO adsorption (and therefore, obviously, also its rate of desorption) is nearly independent of the surface orientation. Even with Pd(110) for which the highest value for the initial adsorption energy was observed E_{ad} drops rapidly to about 35 kcal/mole with increasing coverage[23]. The observed decrease of the reaction rate at higher temperatures is due to the onset of oxygen desorption. Although data on the adsorption energies of oxygen on different Pd surfaces are not available in such detail as for CO, there is certainly again no strong variation with the surface orientation. Thus, it becomes plausible why the oxidation of CO on Pd is a structure-insensitive ("facile") reaction.

III. The Role of Steps

Structural imperfections have been frequently discussed as playing the role of "active centres" in heterogeneous catalysis. Particularly,

100

this was believed to be the case with dislocations which are of decisive importance for the kinetics of crystal growth. Unfortunately, so far there exists no experimental possibility to introduce dislocations with defined densities and structures at clean single crystal surfaces under UHV conditions. However, an alternative possibility for deviations from the perfect lattice structure is the preparation of surfaces with periodic arrays of monoatomic steps. Surfaces of this type are frequently quite stable and can easily be studied by means of LEED[31]. Considerable differences of the reactivities between low index planes and stepped surfaces with reactions involving hydrocarbons were reported by Somorjai, et al.[32]. However, in these cases the surfaces became covered during the reaction by carbonaceous overlayers whose structure and degree of periodicity depended strongly on the presence and type of steps. Thus, one might argue that the variations of the catalytic activity are mainly caused by the structure of the decomposition products which in turn might be influenced (as in normal crystal growth) by the presence of the steps.

The effect of steps on the nucleation and growth of domains of ordered adsorbed layers was recently studied with the system $O/W(110)$[33]: Oxygen adsorption on a W(110) surface causes the formation of two domain orientations of a p(2x1) structure. With the presence of periodic steps along the (111) directions it was observed that one type of domains appeared quite preferentially.

The effect of periodic step arrays on the adsorption energy has been studied for CO[23] and H_2 adsorption[34] on a Pd(111) surface. The results are reproduced in Figs. 9 and 10. For CO the isosteric heat of adsorption, as a function of coverage, is practically identical for both types of surfaces thus supporting the structure-insensitivity of the CO oxidation reaction. However, for hydrogen the initial adsorption energy is higher by about 3 kcal/mole for the stepped surface (thus being similar to the value for the (110)plane) and approaches the data for the low index surface with increasing coverage. This result has to be interpreted in terms of a somewhat higher binding energy of the H-atoms at the adsorption sites near the steps (64.5 kcal/mole instead of 63 kcal/mole).

Quite dramatic effects of the presence of steps were reported for the interaction of hydrogen with Pt(111) surfaces. Somorjai, et al.,[32,36] reported that there is nearly no hydrogen adsorption on low index platinum planes, whereas, this is readily the case with stepped surfaces. These findings made those techniques in "practical" catalysis research questionable for which the surface area of Pt particles is determined from the hydrogen uptake and which have so far been quite successful. Recently Bernasek and Somorjai[35] studied the H_2/D_2 exchange reaction on Pt(111) surfaces with a molecular-beam technique and found that with the low index plane, practically no HD formation could be observed, whereas the stepped surfaces were apparently quite reactive. The conclusion was that the atomic steps play a decisive role in dissociating the H_2 molecules.

Recent results from our laboratory are in contrast to these conclusions[38]: It was found that even at 100 K hydrogen adsorbs dissociatively on a Pt(111) surface with an appreciably high sticking coefficient ($s_o \approx 0.1$) without any indication for the existence of an activation barrier. The isotopic exchange reaction was also observed to take place quite readily in agreement with earlier results of Lu and Rye[39]. However, the adsorption energy was determined to be rather low ($\sim$10 kcal/mole) even at small coverages so that far below room temperature complete desorption takes place after evacuation of the vacuum system. The variation with coverage of the isosteric heat of hydrogen adsorption on a Pt(111) plane, as well as on a stepped Pt(111) surface[40] is reproduced in Fig. 11. Similar to the behaviour on Pd(111) E_{ad} increases for the stepped surface with decreasing coverage to values which are about 3-4 kcal/mole higher. With respect to the strength of the Pt-H bond this corresponds to a variation of only about 3%! A difference of the adsorption energy by 4 kcal/mole, however, is at room temperature equivalent to a variation of the mean residence time by nearly three orders of magnitude. This means that under steady-state conditions a large variation of the surface concentrations occurs and this might explain the pronounced differences observed in the experiments of Bernasek and Somorjai[35] with the activity for the H_2/D_2 exchange reaction. This example nicely demonstrates how under certain conditions even rather small variations of the adsorbate bond strength may drastically effect the catalytic activity so that structural imperfections may indeed play the role of active centres.

It is evident that a quantitative theoretical understanding of such small energetic variations will be rather difficult. An interesting alternative to the naive picture in which the valences of the edge atoms are less saturated and therefore may form a stronger bond with the adsorbate was recently proposed by Ibach[41] in connection with the discussion of the influence of steps on semiconductor surfaces on the kinetics of oxygen adsorption: In an extended study with W(110) surfaces, Wagner and Besocke[42] observed that the presence of steps leads to a lowering of the work function which they interpreted in terms of an early hypothesis of Smoluchowski[43]. According to this model any deviation from a flat surface should cause such an effect since the electron gas does not follow sharp edges. As a consequence, the binding energy of electronegative species should increase mainly for electrostatic reasons.

While this concept probably holds for the interaction between oxygen and semiconductor surfaces, no direct applicability to the examples discussed in the present context may be found: On Pd(111) the adsorption energy of CO is totally unaffected by steps, although the adsorbed CO molecule carries a negative charge. With H_2/Pt(111) steps increase the adsorption energy, although hydrogen adsorption _lowers_ the work function. And finally for the three mostly densed packed clean nickel planes, the work function is reported to vary by 0.3 eV[44] whereas the adsorption energies for hydrogen are nearly equal[21].

103

References

[1] Taylor, H. S., J. Phys. Chem. 30 (1926), 145.

[2] Balandin, A. A., Z. Phys. Chem. B2 (1929), 289; B3 (1929), 167.

[3] Boudart, M., Adv. Catalysis 20 (1969), 153.

[4] van Hardeveld, R., and van Montfoort, A., Surface Sci. 4 (1966), 396.

[5] CRC Handbook of Chemistry and Physics.

[6] Ertl, G., Z. Phys. Chem. N.F. 50 (1966), 46.

[7] Haydock, R., and Kelly, M. J., Surface Sci. 38 (1973), 139.

[8] Feuerbacher, B., Surface Sci. 47 (1975), 115.

[9] Page, P. J., and Williams, P. M., Faraday Disc. 58 (1974), 80.

[10] a) Feuerbacher, B., and Fitton, B., Phys. Rev. Lett. 29 (1972), 786.
    b) Waclawski, B. J., and Plummer, E. W., Phys. Rev. Lett. 29 (1972), 783.

[11] Domke, M., Jähnig, G., and Drechsler, M., Surface Sci. 42 (1974), 389.

[12] Schmidt, L. D., Catalysis Rev. 9 (1974), 115.

[13] Adams, D. L., and Germer, L. H., Surface Sci. 27 (1971), 21.

[14] Delchar, T. A., and Ehrlich, G., J. Chem. Phys. 42 (1965), 2686.

[15] McAllister, J., and Hansen, R. S., J. Chem. Phys. 59 (1973), 414.

[16] Emmett, P. H., in "The physical basis of heterogeneous catalysis" (R. I. Jaffee, ed.), Plenum Press 1976.

[17] Brill, R., Richter, E. L., and Ruch, E., Angew. Chem. Int. Ed. 6 (1967), 882.

[18] a) Dumesic, J. A., Topsoe, H., Khammouma, S., and Boudart, M., J. Catalysis 37 (1975), 503.
    b) Dumesic, J. A., Topsoe, H., and Boudart, M., J. Catalysis 37 (1975), 513.

[19] Ertl, G., Grunze, M., and Weib, M., J. Vac. Sci. Techn., 1976 (in press).

[20] Ertl, G., Grunze, M., and Weib, M., in preparation.

[21] Christmann, K., Schober, O., Ertl, G., and Neumann, M., J. Chem. Phys. 60 (1974), 4528.

[22] Christmann, K., Schober, O., and Ertl, G., J. Chem. Phys. 60 (1974), 4719.

104

[23] Conrad, H., Ertl, G., Koch, J., and Latta, E. E., Surface Sci. 43 (1974), 462.

[24] Gaydon, "Dissociation Energies and Spectra of Diatomic Molecules," London 1953.

[25] Skinner, H. A., Adv. in Organometallic Chemistry (Eds. F. G. A. Stone and R. West), 2 (1964).

[26] a) Tracy, J. C., and Palmberg, P. W., J. Chem. Phys. 51 (1969), 4852.
b) Track, J. C., J. Chem. Phys. 56 (1972), 2736.

[27] Doyen, G., and Ertl, G., Surface Sci. 43 (1974), 197. This treatment was later improved by including also the coupling of the CO 5δ-Orbital to free electron-like metallic s-states, however, without significantly affecting the shape of the calculated energy profiles or the predicted changes of the adsorption energy with surface orientation (G. Doyen and G. Ertl, to be published).

[28] a) Ertl, G., and Rau, P., Surface Sci. 15 (1969), 443.
b) Ertl, G., and Koch, J., in "Adsorption-Desorption Phenomena" (F. Ricca, ed.), Academic Press 1972, p. 345.
c) Ertl, G., and Neumann, M., Z. Phys. Chem. N. F. 60 (1974), 127.

[29] Ertl, G., and Koch, J. Proc. 5th Internat. Congr. on Catalysis, North Holland 1973, p. 969.

[30] a) Matsushima, T., and White, J. M., J. Catalysis 39 (1975), 265.
b) Matsushima, T., Almy, D. B., Foyt, D. C., Close, J. S., and White, J. M., idib., 277.

[31] Lang, B., Joyner, R. W., and Somorjai, G. A., Surface Sci. 30 (1972) 440, 454.

[32] Lang, B., Joyner, R. W., and Somorjai, G. A., J. Catalysis 27 (1972), 405.

[33] Engel, T., von dem Hagen, T., and Bauer, E., to be pulished.

[34] Conrad, H., Ertl, G., and Latta, E. E., Surface Sci. 41 (1974), 435.

[35] Bernasek, S. L., and Somorjai, G. A., J. Chem. Phys. 62 (1975), 3149.

[36] Morgan, A. E., and Somorjai, G. A., Surface Sci. 12 (1968), 405.

[37] Benson, J. E., and Boudart M., J. Catalysis 4 (1965), 704.

[38] Christmann, K., Ertl, G., and Pignet, T., Surface Sci. (in press).

[39] Lu, K. E., and Rye, R. R., Surface Sci. 45 (1974), 677.

[40] Christmann, K., and Ertl, G., to be published.

[41] Ibach, H., Surface Sci. (in press).

[42] Besocke, K., and Wagner, H., Phys. Rev. $\underline{B8}$ (1973), 4597.

[43] Smoluchowski, R., Phys. Rev. $\underline{60}$ (1941), 661.

[44] Maire, G., Anderson, J. R., and Johnson, B. B., Proc. Roy. Soc. $\underline{A320}$ (1970), 227.

Table 1: Initial adsorption energies for hydrogen on Ni single crystal surfaces[21].

Plane	(111)	(100)	(110)
E_{ad}[kcal/mole]	23	23	21.5

Table 2: Adsorption of CO on Nickel. Initial adsorption energy E_{ad} and maximum densities of adsorbed molecules, n_{max}, at T=300 K and $p_{CO} \leq 10^{-4}$ Torr[22].

Plane	(111)	(100)	(110)
E_{ad}[kcal/mole]	26.5	30	30
$n_a \times 10^{15}$[cm^{-2}]	1.1	1.1	1.14

Table 3: Initial adsorption energies for CO on Pd single crystal surfaces[23].

Plane	(111)	(100)	(110)	(210)	(311)
E_{ad}[kcal/mole]	34	36.5	40	35.5	35

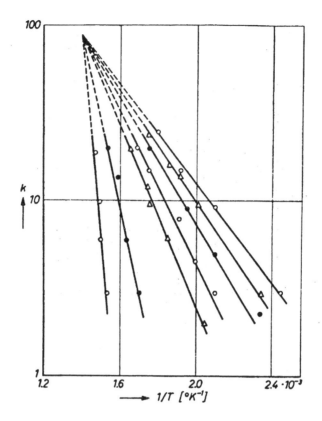

Fig. 1. Arrhenius plots for the rate of N_2O decomposition on crushed
Ge surfaces at different stages of inhibition by adsorbed
oxygen[6].

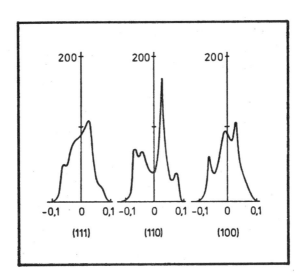

Fig. 2. Local densities of d-states at the (111), (110), and (100)
surface of bcc crystal as calculated by Haydock and Kelly[7].

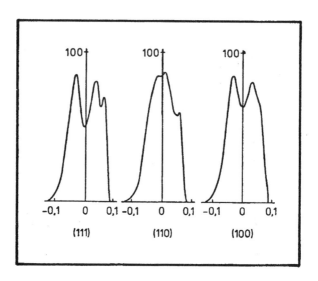

Fig. 3. Local densities of d-states at the (111), (110), and (100)
surface of an fcc crystal[7].

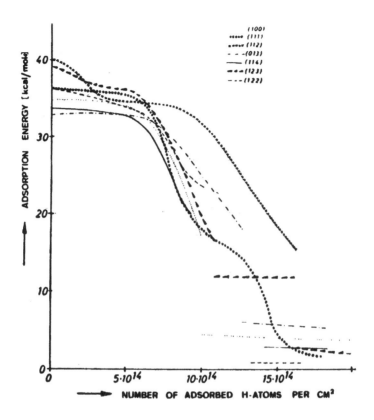

Fig. 4.   Adsorption energy versus coverage for hydrogen on different W
single crystal planes.   (Domke, et al.[11])

111

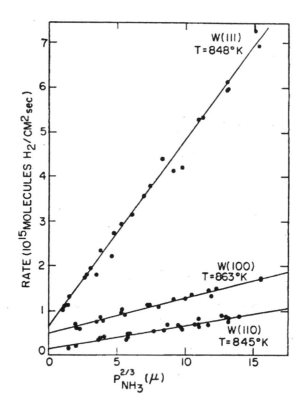

Fig. 5. Rate of NH_3 decomposition as a function of $p_{NH_3}^{2/3}$ on different W single crystal surfaces (McAllister and Hansen[15]).

112

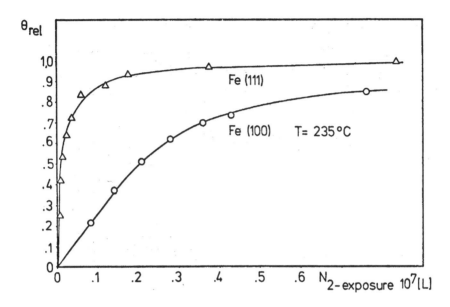

Fig. 6. Adsorption kinetics of nitrogen on Fe(111) and Fe(100) surfaces
at 235°C. Variation of the relative coverage with N_2 exposure
(Ertl, et al.[20]).

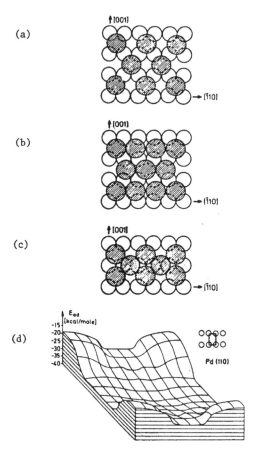

(a)

(b)

(c)

(d)

Fig. 7.   a) – c) Structure models (with increasing coverage) for CO
adsorbed on Pd(110)[23].

d) Theoretical energy profile for the variation of the CO
adsorption energy within the unit cell of the Pd(110) surface[27].

114

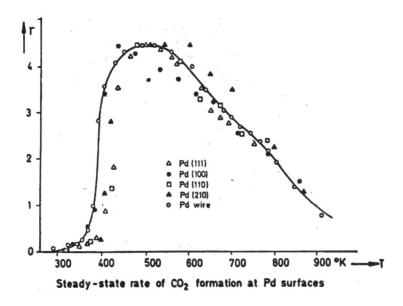

Steady-state rate of CO_2 formation at Pd surfaces

Fig. 8. Steady-state rate of CO_2 formation as a function of temperature on different Pd surfaces. $P_{CO} = P_{O_2} = 10^{-7}$ Torr[29].

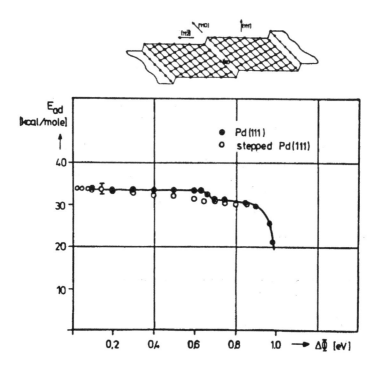

Fig. 9. Isosteric heat of CO adsorption as a function of the work function increase Δφ on a Pd(111) surface (dark circles) and on a stepped Pd(111) surface (open circles). The latter consisted of terraces with (111) orientation, 9 atomic rows in width and separated by monoatomic steps also with (111) orientation[23].

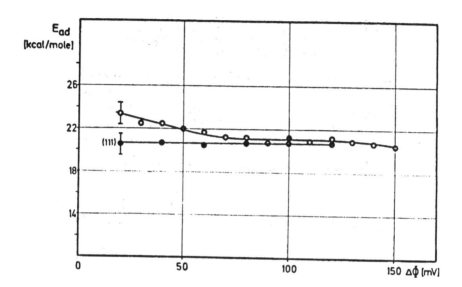

Fig. 10. Isosteric heat of H_2 adsorption as a function of the work function increase $\Delta\phi$ on a Pd(111) surface (dark circles) and on a stepped Pd(111) surface (open circles)[34].

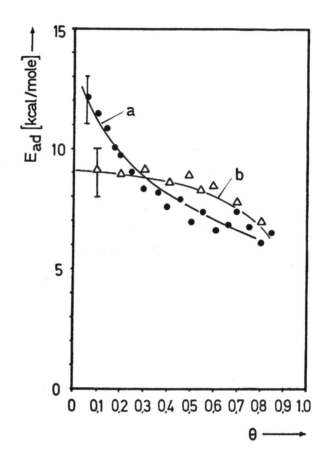

Fig. 11. Adsorption energy as a function of coverage for H_2 on a Pt(111)
surface (curve b, open triangles) and on a stepped Pt(111) surface
(curve a, dark circles)[40].

Part 3. Abbreviation of Comments

Robert S. Hansen

Ames Laboratory, ERDA
Iowa State University
Ames, Iowa

Since my paper with McAllister (J. Chem. Phys. $\underline{59}$, 414-22 (1973))
has been cited by two of the speakers, I would like to add some comments
of my own. Briefly, we found that the rate of ammonia decomposition on
the (111), (100), and (110) faces of tungsten followed the law Rate = A
+ $BP_{NH_3}^{2/3}$, and was independent of nitrogen and hydrogen pressures, over
the temperature range 800 < T < 970 K and pressure range 0.5 < P < 100μ.
Catalytic activities were in the order (111) > (100); > (110); the small
activity observed for (110) was likely due to crystal edges with the
(110) face itself being inactive. For a given face (100) and temperature
the constant A was the same for NH_3 and ND_3, but the constant B was
significantly greater for NH_3. These results imply that two processes
are carrying the decomposition process, represented by the terms A and
$BP_{NH_3}^{2/3}$ in the rate equation. We called these the A and B processes,
respectively. Using the (100) face as a model, we proposed that both
processes occurred on a nearly complete surface WN structure ((1X1)
structure). The WN notation is intended to represent surface stoichiometry only, i.e., one nitrogen atom per tungsten atom. The two
processes suggested were then

$$A \text{ process} \qquad 2WN \overset{k}{\rightarrow} W_2N + 1/2N_2$$

$$B \text{ process} \qquad 3WN + NH_3(g) \overset{K}{\rightleftarrows} W_2N_3H_2 + WNH$$

$$W_2N_3H_2 \overset{k_1}{\rightarrow} W_2N + N_2 + H_2$$

$$2WNH \overset{k_2}{\rightarrow} 2WN + H_2.$$

Presuming the surface fraction of WN to be nearly unity, the fractions
of W_2N, $W_2N_3H_2$ and WNH correspondingly small, the independence of the A
process on reactant and product pressures and the fact that the constant

119

A is the same for NH_3 and ND_3 are accounted for. A steady state treatment of the B process under the same assumptions leads to the $P_{NH_3}^{2/3}$ dependence, and since NH bonds are broken in both rate determining steps the constant B is expected to differ for NH_3 and ND_3 as observed. The (1X1) WN and C(2X2) W_2N structures have been established by LEED; the WNH and $W_2N_3H_2$ structures have not been established (and in the above model would not have sufficiently large concentrations to form recognizable phases).

I would like to use these and related findings as a background to discuss what I think is the outstanding problem in catalysis--the development of a conceptual framework for discussing, in structural detail, the rates of surface reactions, i.e., a conceptual framework for discussing the transition state in such reactions. We generally know the initial reactants and final products of a catalytic reaction, and sometimes have an idea of structures of adsorbed species immediately preceding and immediately following the rate determining step. But we need much better patterns of thought for discussing the rate determining step itself. Let me illustrate the problem with some reactions on the (100) face of tungsten (not because it is catalytically most important, for it surely isn't, but because it is geometrically simple).

Suppose we superimpose a Cartesian coordinate system on W(100) with the unit cell length 3.16Å as length unit and the center of a surface tungsten atom at (o,o), so that each position (m,n), with m,n integers, is located at the center of a surface tungsten atom. The position (1/2, 1/2) is then a hole with 4 tungsten atoms surrounding it in its plane and one below it, so that it is a CN-5 (coordination number 5) position; of course all positions (1/2 + m, 1/2 + n) are similar positions.

Nitrogen adsorbs very readily (sticking coefficient about 0.2) on W(100) until a stoichiometry W_2N is reached, and this produces on annealing the well-defined LEED C(2X2) patterns mentioned several times at this meeting. We believe the nitrogen atoms are in the CN-5 positions, in which case exactly half of these positions are filled. Why does the N_2 sticking coefficient fall several orders of magnitude after the W_2N stoichiometry is reached, when half of the adsorption sites are still vacant?

The bond energy in N_2 is 10 e.V.; it is an extremely strong bond, and can be broken rapidly only if other very strong bonds are <u>in the process of forming</u> as the nitrogen-nitrogen bond is in the process of

120

breaking. Suppose for simplicity that these new bonds are forming in
the CN-5 positions. The empty surface furnishes adjacent pairs of such
positions, so that bonds to both nitrogen atoms can be forming as the
nitrogen-nitrogen bond is breaking. The W_2N C(2X2) configuration is one
in which all CN-5 positions immediately adjacent to an empty CN-5 position
are filled, so only one nitrogen atom in N_2 can be forming a bond in the
transition state which is correspondingly much less favorable. If
nitrogen <u>atoms</u> are provided (by causing $N_2(g) \rightarrow 2N(g)$ by electron bombard-
ment) they are immediately adsorbed until WN stoichiometry (readily
annealed to (1X1) structure) is achieved--in this case new bonds forming
in the transition state do not have to pay a 10 e.V. bond dissociation
price as it has already been paid.

The B process previously outlined for the ammonia decomposition on
tungsten also suggests possible transition states. The W_2N C(2X2)
structure has a lower work function than tungsten, so must have a surface
double layer positive out. Electronegativity considerations indicate
that nitrogen is negatively charged with respect to tungsten; if the
nitrogen bond to the underlying tungsten in the CN-5 position is very
strong the nitrogen adatom can be sufficiently "buried" to account for
the positive out double layer. The work function for WN (1X1) is greater
than that of tungsten, so the nitrogen centers must lie above the plane
of centers of the surface tungsten atoms, which could result simply if
the bond of the nitrogen to the underlying tungsten atom were weaker
than in the W_2N structure as it surely would be. There are of course
other positions which would achieve the double layer negative out result,
but for model purposes let us assume that the nitrogen atoms are still
in the CN-5 position.

The B process then occurs on top of complete nitrogen adlayer, with
a tungsten atom, positively charged, "visible" in the middle of each
elementary square of nitrogen atoms. This tungsten atom is hence function-
ally a Lewis acid, and is an attractive place for the Lewis base ammonia
to sit, bonding to the tungsten through its unshared pair of electrons.
Further, the surrounding nitrogen atoms are negatively charged, and so
are receptive sites for proton transfer. These ideas provide models as
to how the proposed species $W_2N_3H_2$ (which thus really means NH_2^- coordinated
to W^+ in the middle of the elementary square of nitrogen atoms in a WN
structure) and WNH (which thus means H^+ coordinated to N^- in the WN

121

structure). The charges are doubtless incomplete but should be understood as representing bond polarity in each case.

Part 4.   Contribution of John Turkevich,
Department of Chemistry,
Princeton University
Princeton, New Jersey

Palladium catalyst particles (Plate 6067) made from palladium sol
at Princeton Laboratories and examined by ultra high electron microscopy
by Lazlo L. Ban of Petrochemical Research, Cities Service Company. The
magnification on the plate is 2,600,000X. The lattice of the palladium
metal is easily discerned in the polycrystalline particles. Lattice
planes up to the very edge of the particles can be seen on some sides,
though rounded-off amorphous surfaces seem to predominate.

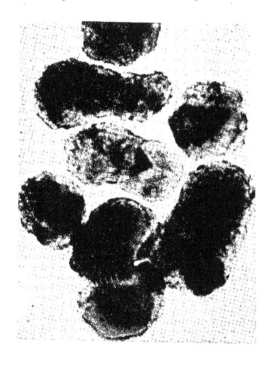

Colloidal gold (175Å diameter plate 3268) prepared at the Princeton Laboratories and examined with high resolution electron microscopy by Lazlo L. Ban of Petrochemical Research, Cities Service Company. The magnification of the plate is 4,000,000X. One millimeter corresponds to 2.5Å. The lattice spacing resolution is about 1.2Å. The spacing of the gold lattice can be seen under the coarser Moiré pattern. In the case of the particle on the bottom of the plate, a spacing of 2.5Å corresponds to the 111 plane and in the particle, second from the top, a spacing of 3.3Å corresponds to 100 plane. The particles themselves are not single crystals but are either multicrystalline (twinned) or have amorphous areas. Spherical shape predominates though there are areas indicating flat surface planes. Flat crystal face appears only on one particle (one face on the particle at the bottom of the plate). In all other cases, the surface is either amorphous or with lattices coming out at an angle. The amorphous nature may be due either to contamination

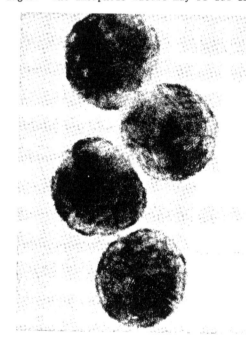

of the surface or to a disordered atomic state of the surface. The appearance of lattice at an angle to the surface and no evidence of lattice parallel to the surface may indicate partial ordering of the gold atoms. The formation of bridges between particles indicates gold atom migration at room temperature after mounting of the particles on the carbon support membrane used in electron microscopy.

The separation of the electronic factor from the geometrical factor
has been in the focus of attention in catalysis. One approach to this
problem is to determine the catalytic activity of alloys. We have
synthesized platinum gold alloy particles in aqueous solution by simultaneous
reduction of gold and platinum chlorides with sodium citrate. The
resultant product in which we varied the platinum to gold ratio was
examined optically. The results obtained by the 50-50% alloy are shown
in the figure (1) together with adsorption spectrum of pure platinum
sol, pure gold sol and a mixture of the two sols. It is seen that the
peak characteristic of gold at 540 nm is absent in the alloy. This is
taken as evidence that the platinum has affected the electronic properties
of gold.

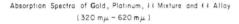

Absorption Spectra of Gold, Platinum, ‖ Mixture and ‖ Alloy
(320 mμ – 620 mμ)

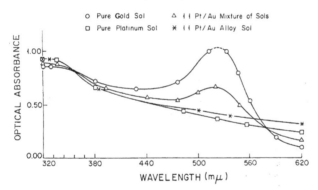

Colloidal platinum (Plate 6064) particles mounted on carbon film in cluster form and examined with high resolution electron microscopy by Lazlo L. Ban of the Petrochemical Research, Cities Service Company. The magnification on the plate is 2,500,000X and one millimeter corresponds to 4.Å. The size of the individual particles of platinum is about 20–30Å. The lattice image of the particle has a spacing of about 3.0Å which may correspond to the 111 plane of platinum. This and the optical spectra of platinum colloid solutions indicate that particles of the platinum as small as 20 to 30Å have metallic properties of bulk platinum.

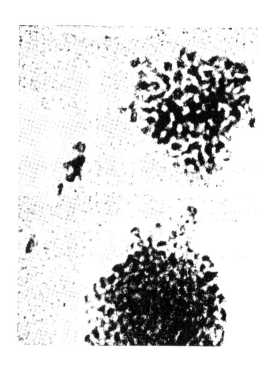

Part 5. The Geometry of Solid Surfaces

C. B. Duke

Xerox Webster Research Center
800 Phillips Road
Webster, New York 14580

A discussion of the atomic geometry of solid surfaces may be divided conveniently into consideration of three topics. How are such structures determined? What is the present state of our knowledge of these structures? What does this information teach us about surface chemistry? In this presentation each of these subjects is examined, in turn.

Two techniques commonly are used for the determination of the atomic geometry of the surfaces of crystalline solids. Field-ion microscopy consists of the imaging on a hemispherical screen of ions generated in the vicinity of a small tip (a few thousand angstroms in radius) following field evaporation to produce a suitable tip surface[1,2]. In this fashion an image of the surface is produced on the screen, suitable for photographing. This technique is discussed and such photographs are displayed in this session by Dr. A. J. Melmed. The second widely-employed structure-analysis technique is low-energy electron diffraction ("LEED"). In particular, the analysis of the configuration of beams of electrons elastically reflected from the surface of a crystalline solid can be analyzed to extract the translational symmetry parallel to a planar surface, whereas the intensities of these beams must be examined in order to determine the atomic geometry of the surface[3,4]. These experiments are sensitive to <u>surface</u> structure because the strong interactions of such low-energy ($5 \text{ eV} \leqslant E \leqslant 500 \text{ eV}$) electrons with the constituents of solids requires that for their elastic emission from a solid, they must emanate from a depth of, at most, about 10Å from its surface. The details of these interactions and their consequences have been described in the literature[4-6]. Here, it is appropriate only to observe that these strong interactions are responsible for the surface sensitivity of electron scattering and emission spectroscopies (e.g., LEED, photoemission and Auger-electron-emission), and that care in interpreting such spectra must be exercised for exit electron energies $E \gtrsim 200 \text{ eV}$, in which case they begin to reflect the bulk as well as surface properties of the sample.

127

A number of reviews of the status of surface-structure determination via LEED intensity analysis have appeared recently[3,4,6-10]. The general trend which emerges both from these reviews and from more recent results (especially for semiconductors[11-13]) is an intimate relationship between the nature of the chemical bonding of a bulk solid and the structure of its clean low-index surfaces (e.g., cleavage planes). Most metals exhibit surface geometries essentially identical to those in the bulk with the possibility of a small (i.e., 10% or less) contraction of the uppermost lattice spacing on more open (higher index) surfaces. Homopolar semiconductors (Si and Ge) are characterized by translational-symmetry-breaking atomic rearrangements even on their (111) cleavage planes, apparently driven by the tendency of "dangling" covalent bonds to yield insulating rather than metallic behavior parallel to the surface ("Peierls instabilities"). The cleavage planes of heteropolar semiconductors [the (110) plane for zincblende and (1010) (11$\bar{2}$0) planes for wurtzite geometries] exhibit the same translational symmetries as their bulk counterparts. Subtle bond-length-conserving rotations of the uppermost atoms may occur, however, giving a rippled appearance to the surface much like waves on a choppy sea. These result from the competition between ionic and covalent contributions to the surface energy. Such competition is not relevant for the polar zincblende (111) and wurtzite (0001) and (000$\bar{1}$) surfaces, which are less stable and tend to exhibit contractions of the upper layer spacing on the cation but not anion faces (due to the presence of lone-pair electrons on the anions but not the cations[13]). Little is known about the surface structure of molecular solids, alkali halides, and transition-metal oxides. The cleavage faces of transition-metal layer dichalcogenides are thought to exhibit surface structures identical to the corresponding bulk structure[14]. All of the low-index surfaces of the transition metal oxides of interest for catalysis are amenable to structure determination via LEED intensity analysis. Thus, only the absence of the requisite intensity data precludes the determination of these important structures.

Analyses of the structures of adsorbed overlayers are in a more primitive state than those of clean low-index surfaces. Although some controversy has occurred in the literature, the structure of the most extensively examined system, Ni(100)-C(2x2)-S, is now agreed upon by all the groups which examined it[15]. In spite of the small number of systems studied[7] and the uncertainties in the determined structures, however, one trend seems to be emerging. The adsorption of even reactive gases (e.g., 0) on the low-index faces of fcc metals (e.g., Ni) does not seem to lead to the formation of surface compounds (e.g., NiO) except under high temperatures or pressures. Thus, the adsorbed atoms seem to occupy the hollows of the metal surfaces without substantial distortion of the metal substrate to a far greater extent that anticipated by early workers.

Whereas it is commonly supposed that such ultra-high-vacuum clean-surface and low-coverage adsorbate structures have little relevance to the surface chemistry of "practical" catalysts, such is not the case in the semiconductor electronics industry. Indeed, it is well-known[13] that surface strains on the cation faces of III-V crystals preclude crystal growth, inhibit mechanical damage, and reduce solution oxidation rates. Obviously, such structure-property relationships are highly useful in the processing of semiconductors. Moreover, recently Rowe et al.[16] have proposed an intimate relationship between surface structure and the formation of rectifying metal-semiconductor contacts. Consequently, while the gloomy prognosis for the impact of ultra-high vacuum surface-structure work on catalytic chemistry advanced by many speakers at this symposium may well prove correct[17], it certainly will not mitigate the substantial importance of such studies for materials processing in the electronics and electrophotographic industries. Perhaps a more fruitful approach even in the area of catalysis (in which kinetics are probably defect-dominated and hence ill suited for direct study via structure determination) may be the discernment and exploitation of structure-property relationships analogous to those which have proven so valuable in semiconductor materials science.

References

[1]   Müller, E. W., in Proceedings of the International School of Physics 'Enrico Fermi', Course LVIII. (Enditrice-Compositori, Bologna, 1975), p. 23.

[2]   Müller, E. W., and Tsong, T. T., Field Ion Microscopy (American Elsevier, New York, 1969).

129

[3]   Duke, C. B., in Proceedings of the International School of Physics,
        'Enrico Fermi', Course LVIII. (Enditrice Compositori, Bologna, 1975),
        pp. 99, 174.

[4]   Duke, C. B., Adv. Chem. Phys. 27, 1 (1974).

[5]   Duke, C. B., Crit. Rev. Solid State Sci. 4, 371 (1974).

[6]   Duke, C. B., in Proceedings of the International School of Physics
        'Enrico Fermi', Course LVIII. (Enditrice Compositori, Bologna,
        1975), p. 52.

[7]   Duke, C. B., Japan. J. Appl. Phys. Suppl. 2, 641 (1974).

[8]   Duke, C. B., Lipari, N. O., and Laramore, G. E., Nuovo Cimento
        23B, 241 (1974).

[9]   Lagally, M. G., Bucholz, J. C., and Wang, G. C., J. Vac. Sci. Technol.
        12, 213 (1975).

[10]   Strozier, J. A., Jr., Jepsen, D. W., and Jona, F., in Surface Physics
        of Materials, J. M. Blakely, Ed., (Academic, New York, 1975), Vol. I,
        p. 2.

[11]   Duke, C. B., and Lubinsky, A. R., Surface Sci. 50, 605 (1975).

[12]   Lubinsky, A. R., Duke, C. B., Chang, S. C., Lee, B. W., and Mark, P.,
        J. Va. Sci. Technol., Jan/Feb. (1976).

[13]   Gatos, H. C., J. Electrochem. 122, 287C (1975).

[14]   Tong, S. Y., and Van Hove, M., private communication.

[15]   Duke, C. B., Lipari, N. O., and Laramore, G. E., J. Vac. Sci.
        Technol. 12, 222 (1975).

[16]   Rowe, J. E., Christman, S. B., and Margaritondo, G., Phys. Rev.
        Lett. 35, 1471 (1975).

[17]   Duke, C. B., Crit. Rev. Solid State Sci. 4, 541 (1974).

Part 6. Contribution of A. J. Melmed

National Bureau of Standards,
Washington, DC   20234

One experimental technique which can provide structural information, expecially surface structural information about particles in the size range of some practical catalysts is field-ion microscopy[1] (FIM). To my knowledge FIM has not yet been applied to actual catalyst particles, but in principle it is possible.

I will address just one aspect of FIM because I believe that it is relevant to some natural confusion that has crept into the minds of those who think about catalysts in terms of atomic surface geometry. Most field-ion micrographs published are images of surfaces which have been prepared, in the final stage, by low-temperature field-evaporation[1]. This process results in surfaces which look similar to what one would expect on the basis of constructions using hard spheres to represent atoms. However, this process is very artificial in terms of processes occurring outside the field-ion microscope. Thus, the near-ideal structures produced by field-evaporation do not occur as the result of ordinary annealing, for example.

The surface atomic structures of thermally annealed platinum, iridium, and tungsten imaged by FIM clearly showed a large degree of thermal disorder. An example of disorder introduced mechanically in an iridium specimen was also shown. The intended message: Be aware of atomic surface disorder, as well as atomic order, which may well be present on the surfaces of real catalysts.

[1]   E. W. Müller and T. T. Tsong, Field Ion Microscopy; Principles and Applications (Elsevier, New York, 1969).

131

Part 7. General Discussion for the Session

A. B. Anderson

Chemistry Department
Yale University,
New Haven, CT

A. B. Anderson: I have a question directed to Professor Ertl. You have
shown calculated electronic densities of states for various crystal
faces of fcc and bcc metals. It seemed that in one case some kind of
correlation with surface properties may have existed and in the other
case there was apparently no correlation. Can a case be made for the
pertinence of detailed electronic density of states determinations to
problems in chemisorption and catalysis? I have seen these density of
states plots presented in various places, but sometimes without being a
part of an argument. Now I am not against exploring the possible implica-
tions of densities of states for surface events, but until we know what
bond shapes really mean to a particular reaction on a surface[1], their
presentation, merely because they are something calculatable, seems glib.

Frequently, adsorbate levels such as sigma and pi energy levels in
hydrocarbons lie several electron volts beneath the bottom of transition
metal s-d bonds. When these orbitals interact with bond orbitals they
may not be sensitive, as far as some properties of the interaction are
concerned, to details of the s-d bonds. We find adsorbed molecules have
an ability to induce a strong interaction with a metal surface because
of the high density of atomic d-orbitals in the surface which find ways,
by appropriate linear combinations, to form strong bonds for various
adsorbate locations above various surfaces[2,3].

So my question is: Can a case be made for the pertinence of detailed
density of states calculations to our present understanding of chemisorption
and catalysis or has one been made?

Professor Ertl: I am in agreement with what you say and believe the
results in my talk did not show any such pertinence. However, sometimes
adsorbate levels may be in the s-d bond.

132

A. B. Anderson: Yes, as in the case of π* orbitals in some instances. I have a comment and a question directed to Professor Turkevich. You have shown how microcrystals of alumina supported platinum appear to have bulk structures inside but apparently a random structure at and near the surface. I have calculated binding energies and structures for two to six atom clusters of tin, titanium, chromium, iron, and nickel atoms[3] and the structures have no resemblance to the bulk structures. This seems to be in agreement with your observations. Now I cannot be entirely certain of the transition metal calculations, as there are no experimental binding energies for comparison but in the case of tin, Gingerich and coworkers at Texas A and M University[4] have found by experimental methods, the binding energy per atom for these clusters and my calculations agree within around a kilocalorie per mole or two. The preferred structures were found by varying the coordinates of all the atoms to find the lowest energy. The structures were quite independent of the bulk diamond-like structure; for example, the five atom cluster took the form of a trigonal bipyramid. Because of the accuracy of the energies, it seems likely that the calculated structures are correct.

And so my question has to do with the temperature of your platinum microparticles, the possibility of metastability and our understanding of the nucleation process.

Professor Turkevich: The particles are in structural equilibrium.

A. B. Anderson: Perhaps some interesting things are going on here. Perhaps the nucleation process involves condensation of tiny crystallites with various structures, or perhaps one atom adds at a time. Frequently, according to the calculations, for a particular number of atoms in a microcrystal, a unique structure is preferred and will be preferred up to quite high temperatures. But as condensation occurs there is a transition to the bulk structure in the inner part of the particle. At least, this is what your work suggests to me.

133

References

[1] Some efforts are being made for generalized interactions between model surface densities of states and model adsorbates. See J. W. Gadzuk, Surface Science 43, 44 (1974), M. J. Kelly, J. Phys. C.: Solid State Phys. 7, L157 (1974), and references in these papers to work by T. L. Einstein, J. R. Schrieffer, T. B. Grimley, D. M. Newns and others.

[2] Anderson, A. B., and Hoffmann, R., J. Chem. Phys. 61, 4545 (1974).

[3] See works in footnotes to the talks by A. B. Anderson at this conference.

[4] Gingerich, K. A., Desideri, A., and Cocke, D. L., J. Chem. Phys. 62, 731 (1975).

Comments by Arthur Wm. Aldag, Jr.

School of Chemical Engineering and Materials Science
University of Oklahoma
Norman, Oklahoma

We have been engaged in an experimental study[1] of the decomposition
of formic acid on a single crystal nickel (100) McCarthy, et al.[2] have
studied the nickel (110) surface. Both studies employ the flash desorption
technique where formic acid is absorbed at room temperature and the
evolution of the products followed mass spectrometrically as the crystal
is heated. On both the (100) and (110) surfaces H_2 and CO_2 are liberated
first with the two peaks superimposed on each other. At a higher temperature, CO is liberated leaving a surface oxide that can be detected by
Auger Spectroscopy. A study of the interaction of the reaction products
H_2, CO and CO_2, alone on each surface indicates differences in the
binding energies and sticking coefficients not uncommon to many other
studies of simple absorbates on well defined metal surfaces. However
there is a pronounced difference in the formic acid decomposition kinetics
on Ni (100) and Ni (110). Madix finds that the H (and CO_2) appears in a
remarkably sharp flash peak with a half-width of about 6° K. The only
plausible explanation is that the decomposition on Ni (110) is autocatalytic leading to a "kinetic explosion." By contrast, we find that
on Ni (100), H_2 and CO_2 peaks are of "normal" half-width and appear
to obey 2nd order decomposition kinetics. This would appear at face
value, to be a rather striking example of the structure factor referred
to earlier. Possibly the less dense (110) surface offers a more favorable
environment for propagation of the branching chain. We also find that
there is a small (2%) amount of residual oxygen on the "clean" Ni (100)
surface which cannot be removed. An alternate explanation might be that
this oxygen serves to terminate the branching step on Ni (100). This
alternative would then fall more under the heading of the "ligand factor"
referred to by Professor Boudart.

References

[1] Lee, C-O, Peavey, J. H. and Aldag, A. W., Submitted to J. of Catal.

[2] McCarthy, J., Falconer, J., Madix, R. J., J. of Catal., 30, 235 (1973).

Bonding Properties of Stepped Transition Metal Surfaces

G. S. Painter[1]

Metals and Ceramics Division

Oak Ridge National Laboratory[2]

Oak Ridge, Tennessee 37830

R. O. Jones

Institut für Festkörperforschung der Kernforschungsanlage

Jülich, 517 Jülich, Germany

P. J. Jennings

School of Mathematical and Physical Sciences

Murdoch University, Murdoch, WA 6153 Australia

A great deal of interest has been shown recently in the use
of theoretical clusters models[1] to determine: (1) how electrons
are distributed spatially and energetically in small crystallites
and at surfaces[2-6], (2) how electrons respond to surface disorder
such as steps and kinks[7,8], and (3) how electronic characteristics
are related to chemically active sites on surfaces[7-10]. Much of
this interest has been stimulated by experimental observations[11] of
large differences in surface phenomena associated with prepared smooth
(low index) surfaces as compared with corresponding behavior for typical
disordered surfaces. The origin of structure insensitivity in certain
reactions has been phenomenologically developed and discussed recently by
Boudart[12]. Surface cluster models offer new insight into these problems

---

[1]Guest scientist with the Institut für Festkörperforschung der
Kernforschungsanlage Jülich on exchange from ORNL (1974-75).

[2]ORNL is operated for the Energy Research and Development Administration
by the Union Carbide Corporation.

of surface interactions since they provide detailed information about the electronic energy distribution and local bonding characteristics at the surface. The basic formulation of the various cluster theories involves interactions among a finite number of atoms, thus these models form a useful complement to those approaches based on the extended nature of the substrate[13-15].

As part of a series of studies of surface bonding characteristics, we have carried out calculations of the electronic structure of clusters chosen to simulate simple stepped transition metal surfaces in the absence of an adsorbate. Some interesting features of the orbital density near the step region were recently reported[7,8], and a possible relation to observed enhanced activities of small particle and stepped surface catalysts was suggested. In connection with the conference topic "geometrical effects" we wish to briefly summarize some of these results and discuss the origin of the various orbital features which occur at simple surface irregularities. The origin and general occurrence of these features near edges and corners suggest that similar behavior should be expected for more extended and complicated types of disorder as would be found for actual surfaces.

For simplicity we chose a thirteen atom cluster with atoms positioned to simulate a simple step on the (100) face of a transition metal surface. Specifically, we chose nine atoms in a square arrangement 2A on a side where A = the near-neighbor separation in the plane, with four atoms in interstitial sites a distance h above the plane where h = A/2 for bcc and h = A/√2 for fcc symmetries. The cluster symmetry is thus $C_{4\Delta}$. A one-electron model hamiltonian was used with Slater's statistical exchange approximation, and the potential function average to "muffin-tin" form which is spherically symmetric inside touching spheres centered on each atom and outside a sphere surrounding the cluster of atoms, and constant in the volume between the spheres. Past experience with this model suggests that the muffin-tin approximation is in general too restrictive for treating the energetics of chemisorption, however for qualitative and comparative studies the simplicity of the model offers attractive advantages over alternative methods. In the present scattered-wave model the simple form of the secular matrix[16] allows us to include more atoms in the cluster than would otherwise be possible, and this

137

aspect of the calculation was deemed more important for our consideration
than say the inclusion of the non-muffin-tin corrections to the potential.
The calculation was carried out with programs constructed for solving
the multiple-scattering equations for the bound electron states in a
system of muffin-tin scatterers and is described in more detail elsewhere[8].
For the present discussion, the results to be presented are not very
sensitive to the particular details of the technique or parameters of
the model. The features emphasized here should emerge as characteristics
of the system in either a scattered wave or an LCAO approach.

The eigenvalue spectra and histogram densities-of-states for the
clusters of the first transition metal series show some interesting
similarities with corresponding features of the bulk solid. In Fig. 1
we compare our histogram density-of-states for a cluster of thirteen
iron atoms (top panel) with the band structure results of Wood[17] for
bcc iron (bottom panel). The cluster distribution has been shifted to
align the Fermi energies (dashed lines) of the two curves, and note that
the energy scales are not the same. The reduced number of interactions
among neighboring atoms in a small cluster naturally leads to a set of
d-levels whose breadth is not as great as that obtained in the bulk
limit. In addition the different boundary conditions for the cluster as
compared to the extended crystal lead to different band widths. In Fig.
1 the "band" of cluster states constructed by summing energy-normalized
Gaussians at each eigenvalue, is somewhat greater than half the width of
the bulk d-bands. Corresponding to this reduction is an overall increase
in the density-of-states of the cluster over the energy range including
the d-levels, since the total number of states per atom is the same in
each case. Thus the density-of-states at the Fermi energy (highest
occupied level) is 5.90 electrons per atom per eV for the cluster compared
with 3.49 for the bulk case. The distribution of discrete levels contributing
to the density-of-states of the cluster is given in Fig. 1 with the
various $C_{4\Delta}$ symmetry representations noted at the left.

In considering the relationship between clean substrate characteristics
and general surface activity, the calculated narrowing and enchancement
of the density-of-states in small clusters may be significant as far as
a density-of-states factor is concerned. Surely one of the most noticeable

138

aspects of the state density for small metal clusters is the similarity observed in the shape of the density-of-states to that for the extended solid, particularly at the top of the band[4,7,8]. The main structural features that appear in the bulk limit, particularly the position of the Fermi level in the main peak, have emerged even with this limited number of atoms. Apparently the principal influence of including more neighbors in this cluster model is a broadening of the band and an attenuation of the peaks with no great alteration of the main structure. To the extent that the results for this finite cluster represent the local density of states at the surface of the extended solid, the similarities with the bulk may explain instances in which correlations of surface activity with bulk density-of-states have been observed[18].

One primary motivation for using the cluster surface molecule approach to study surface interactions is that it offers the advantages inherent in molecular orbital methods for extracting information about the nature of the bonding in the system. A knowledge of surface bonding is important not only for a quantitative description of the reordering of d-levels at clean transition metal surfaces but also for a qualitative understanding of precursor and intermediate states in chemisorption and molecular dissociation. Some features which are rather characteristic of the orbitals calculated for the surface clusters of our studies will be discussed in the following.

It is of course incomplete to discuss surface bonding properties without specifying the adsorbate since the respective energy spectra of the surface cluster and adsorbate determine the interaction. However, the complexity of a given composite system, and the existence of the vast number of possible reaction pathways have led to searches for establishing less precise but more general concepts of surface activity through correlations between properties of the isolated adsorbate and clean metal substrate. Discussions in terms of bonding have often involved models utilizing the free atomic d-orbital properties or metal bond character of the bulk[19]. Most of our discussion will concern calculated cluster eigenfunctions in the absence of an adsorbate, and emphasis will be placed on the role that bonding among the surface atoms plays in determing the bonding properties of the surface with an adsorbate.

139

In particular, our interest is concerned with the influence of surface
geometrical disorder on the bonding properties, for example, in the
vicinity of a step.

In Fig. 2 we show contour plots of the orbital density, $|\psi(\vec{r})|^2$,
for an eigenstate which lies near the Fermi level in vanadium, and is
representative of one type of solution of the B_2 symmetry species.
The plane of the plot passes through the four atoms comprising the step
in (a) and in (b) the plane is parallel to that in (a) but h/2 above it,
where "h" is the atomic step height. The density magnitudes corresponding
to adjacent contours differ by a factor of two. The sign shown in each
quadrant is that associated with the wavefunction, $\psi(\vec{r})$, in the x > 0,
y > 0 quadrant. Within the latter quadrant, the sign changes of
$c(\vec{r})$ are unspecified. The density in the plane through the step atoms
clearly shows antibonding between partial waves with large components of
d_{xy} symmetry on each step site. A function $\chi = A(\phi_1 + \phi_2 + \phi_3 + \phi_4)$,
where ϕ_i is a d_{xy} orbital on the i_{th} step atom belongs to representation
B_2 and forms an anti-bonding component of the wavefunction. The near-
neighbor step atoms form ddπ symmetry bonds through the destructive
superposition of the partial waves in the overlap region between adjacent
sites. Corresponding to this reduction in density in the region between
the sites, the density is shifted to the exterior lobes, i.e., this
state is associated with the formation of charge lobes at the step
corners. In the plane above the step layer (Fig. 2b) similar behavior
occurs originating mainly in this region from a B_2 basis function formed
from step side d_{xz} and d_{yz} orbitals. Specifically, the major B_2 component
in this region can be written as a combination of d_{xz} and d_{yz} orbitals
drawn from each step site such that each orbital is anti-bonding with
each near-neighbor orbital through either ddπ or ddδ interactions. This
results in a reduced density along the z-axis (although there is an
important density contribution in the hole centered location) with a
compensating increase in density in the lobes extending away from the
corners. Thus the origin of the extended lobe characteristics of the
electron distribution for this state is simply the anti-bonding interactions
among t_{2g} orbitals (d_{xy}, d_{xz}, d_{yz}) on neighboring step sites subject to
the orthonormality constraint on the state which shifts charge out of
the overlap region. Although the features of this orbital density over
the step are rather simple to analyze, this is not generally the case,

140

as illustrated in Fig. 3 for a plane midway between the step layer and the layer of nine atoms for a state in iron. Also, the density in a given plane for a given symmetry type can change significantly with energy—although an atomic orbital analysis assists in understanding the interactions which determine a given state, it can be misleading to describe the nature of the state from postulated interactions among an assumed set of atomic orbitals.

The symmetry characteristics of the various cluster orbitals provide information about the types of bonding allowed for adsorbates at different sites of the cluster, based on rules for the conservation of orbital symmetry[20] in a reaction. A knowledge of the distribution of substrate orbital density makes it possible to qualitatively describe the relative bond strengths for an absorbate at various sites, to the extent that simple orbital overlap plays a role in the bonding. The orbital distributions of symmetry type B_2 shown in Fig. 2 illustrate these points in a simple way. Consider an identical atom (vanadium) approaching the cluster along the z-axis. A d_{xy} orbital on the z-axis belongs to the B_2 representation and can form a ddδ bonding configuration with the step orbital distribution, which is made up of d_{xy}, d_{xz} and d_{yz} single site contributions and has d_{xy} symmetry about the z-axis itself. This mechanism for the formation of a bond at a (001) hole-centered site in fact is operative in the bonding of the V atom at the origin of the cluster with the step atoms above. Furthermore, a position on the z-axis over the step atoms would be the stable bonding site for a V atom for growth of the cluster to form the crystalline solid.

The cluster wavefunctions for the totally symmetric A_1 representation are significant for σ-bonding in the hole-centered (001) site (Fig. 4). The partial wave contributions from the step sites are of d_{z^2} symmetry and are summed in the A_1 representation to produce a totally symmetric contribution to the orbital density which is concentrated over the step atoms. In contrast to the B_2 representation, the d_{xz}, d_{yz} step orbitals are involved in the formation of bonding ddπ, ddδ combinations between near-neighbors in the step plane, thus enhancing the density about the z-axis and producing a state conducive for σ-bonding at a (001) hole-centered site on the step (Fig. 4b).

141

The importance of the directionality of the d_{xz}, d_{yz} step orbital
contributions in σ-bonding of an adsorbate can be illustrated in the
case of CO chemisorbed at the (001) hole site of a Ni_{13} cluster. In
Fig. 5, we show contours in the x = y plane for a σ-bonding eigenstate
in the Ni_{13}-CO system. The plane of the figure passes through nickel
atoms at the center (left) and corner (right) of the layer of nine;
through one of the step atoms above, and the CO is along the z-axis with
the C atom located nearer the Ni cluster. The d_{xz} and d_{yz} partial wave
components of the "substrate" (particularly the step sites) admix with
other orbitals to give a hybrid directed toward the CO, forming a good
bond with the CO molecular orbital formed from s-p_z combinations. In
Fig. 6, we show density contours for this state in a plane passing
parallel to the step layer and somewhat more than h/2 above it (the
plane is actually $\ell/2$ below the C atom where ℓ = 2.13234 a.u., the CO
bond length). Clearly, a significant amount of charge is associated
with this σ-bond, and involves the metal electron distribution which
initially was concentrated over the step layer similarly to that shown
in Fig. 4b.

In conclusion, we note that the great complexities of most systems
of practical concern in catalysis make it unfeasible to apply cluster
models directly to the problems of interest. However, the fundamental
information obtained in simpler systems and processes, e.g., chemisorption,
should play a useful role in the interpretation of processes involved in
complex catalytic reactions. A large amount of spectral data from
various surface probe experiments performed on adsorbate covered surfaces
now exists. Theoretical attention to this problem has grown
recently[5,9,10,21-25], and it appears that a productive interplay
between theory and experiment has emerged in relating calculated and
measured spectral features. In order to study the energetics of intermediate
products formation and molecular dissociation at surfaces, various
refinements of the present models are required, and efforts to implement
these are currently in progress. Within the present model, the cluster
solutions may be useful as a starting point for simplified calculations
of the energetics of adsorption and dissociation, as for example, in the
perturbation approach employed by Deuss and van der Avoird[26]. In the
near future however, we anticipate that the greatest use of the cluster

142

models will involve interaction with experiment to study bonding configurations and stable adsorption sites on surfaces through correlations of surface spectral features.

G.S.P. and P.J.J. would like to express their appreciation to Professor G. Eilenberger and the members of Theorie I for their interest in this work and the hospitality extended to us during the period of our visit with the Institute when most of this work was carried out.

References

[1] See, for example, review by T. B. Grimley in Dynamical Aspects of Surface Physics, Varenna Lectures (1974).

[2] Baetzold, R. C., and Mack, R. E., J. Chem. Phys. 62, 1513 (1975) and references therein.

[3] Johnson, K. H., and Messmer, R. P., J. Vac. Sci. Technol. 11, 236 (1974).

[4] Messmer, R. P., Tucker, C. W., Jr., and Johnson, K. H., Chem. Phys. Letters 36, 423 (1975).

[5] Fassaert, D. J. M., Verbeek, H., and van der Avoird, A., Surface Sci. 29, 501 (1972).

[6] Gadzuk, J. W., in Surface Physics of Crystalline Materials, edited by Blakely, J. M., (Academic Press, N.Y., 1974).

[7] Painter, G. S., Jennings, P. J., and Jones, R. O., J. Phys. C: Solid State Phys. 8, L199 (1975).

[8] Jones, R. O., Jennings, P. J., and Painter, G. S., Surface Sci. (in press).

[9] Anderson, A. B., and Hoffman, R., J. Chem. Phys. 61, 4545 (1974).

[10] Baerends, E. J., Ellis, D. E., and Ros, P., Theoret. Chim. Acta (Berl.) 27, 339 (1972).

[11] Somorjai, G. A., and Blakely, D. W., Nature 258, 580 (1975) and references therein.

[12] Boudart, M., J. Vac. Sci. Technol. 12, 329 (1975).

[13] Smith, J. R., in Interactions on Metal Surfaces, edited by R. Gomer (Springer-Verlag, Berlin, 1975) and references therein.

[14] Caruthers, E. B., and Kleinman, L., Phys. Rev. Lett. 35, 738 (1975) and references therein.

[15] Kesmodel, L. L., and Falicov, L. M., Solid State Comm. 16, 1201 (1975).

[16] See, for example, K. H. Johnson, Adv. Quant. Chem. 7, 143 (1973).

[17] Wood, J. H., Phys. Rev. 126, 517 (1962).

[18] Dowden, D. A., in 1975 Proc. Trieste Winter College on Surface Science, IAEA, Wien (to be published).

[19] Sinfelt, J. H., CRC Critical Review of Solid State Sciences 1, 311 (1974).

[20] Woodward, R. B., and Hoffman, R., The Conservation of Orbital Symmetry (academic Press, New York, 1970).

[21] Niemczyk, S. J., J. Vac. Sci. Technol. 12, 246 (1975).

[22] Batra, I. P., and O. Robaux, J. Vac. Sci. 49, 653 (1975).

[23] Anders, L. W., Hansen, R. S., and Bartell, L. S., J. Chem. Phys. 62, 1641 (1975).

[24] Lang, N. D., and Williams, A. R., Phys. Rev. Lett. 34, 531 (1975).

[25] Harris, J., and Painter, G. S., Phys. Rev. Lett. 36, 151 (1976).

[26] Deuss, H., and van der Avoird, A., Phys. Rev. B8, 2441 (1973).

143

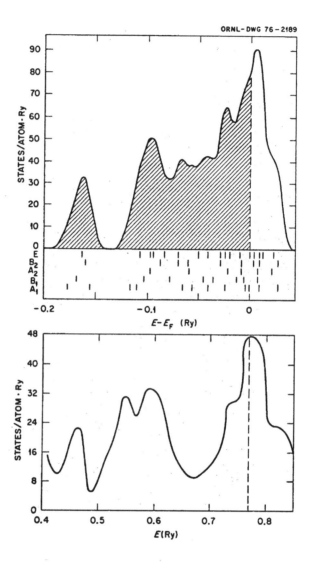

Figure 1.   Comparison of the densities-of-states for iron calculated
from (top panel) the discrete levels of a thirteen atom cluster
and (bottom panel) band structure of bcc crystalline iron by
J. W. Wood. Distributions have been shifted to align the Fermi
levels; note that energy scales differ by a factor of nearly 2.

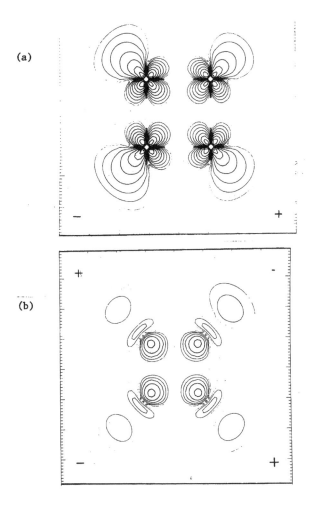

Figure 2. Charge density contours from a B_2 symmetry orbital of a vanadium thirteen atom cluster in planes (a) through the four step atoms ($z = h$) (b) above the step ($z = 3h/2$); h = step height. Adjacent contours differ by a factor of two; signs in each quadrant denote symmetry of orbital.

Figure 3. Charge density contours for an A_1 symmetry orbital of an iron thirteen atom cluster in a plane ($z = h/2$) midway between the nine atoms of the "flat" surface ($z = 0$) and the four step atoms ($z = h$).

146

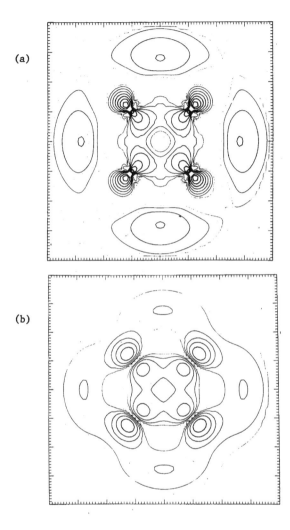

Figure 4. Charge density contours for an A_1 symmetry orbital of the vanadium cluster in planes (a) z = h and (b) z = 3h/2 as described in Fig. 2.

147

Figure 5. Orbital charge density contours for an A_1 symmetry state
in a Ni_{13}-CO cluster. The plane of the figure (x = y) passes
through the central and a corner Ni atom of the "flat" surface,
a Ni step atom above and the CO molecule in the (001) hole-
centered site with the carbon atom down.

Figure 6. Charge density of the A_1 orbital of Fig. 5 in a plane parallel
to that containing the four step atoms and $\ell/2$ below the C atom
where ℓ is the CO bond length.

Part 8. Recorder's Summarizing Comment

The discussion encompassed an assortment of geometrical considerations.
Experimental methods for determining atomic surface geometry were briefly
reviewed, and bond geometry at surface defects was theorized. Geometric
specificity in chemisorption was discussed at length and questions about
geometric specificity in some catalytic reactions were pondered. The
overall implication clearly seemed to be an awareness of a great lack of
knowledge about the geometrical details of real catalysts. To the
extent that this gap was realized by the participants, this session
succeeded.

ELECTRONIC
STRUCTURE

Moderator:

R.E.WATSON

Brookhaven National Laboratory

Lecture by:

J.W.GADZUK

NBS

Session 4.

Electronic Structure

Panel Members:

A. B. Anderson,
Yale Univ.

J. W. Gadzuk,
NBS

D. R. Hamann,
Bell Tel.

R. P. Messmer,
Gen. Elec.

J. R. Smith,
Gen. Motors

Recorder:

J. W. Gadzuk,
NBS

PANEL DISCUSSION: ELECTRONIC STRUCTURE

Chairman: Dr. R. E. Watson, Brookhaven National Laboratory
Recorder: Dr. J. W. Gadzuk, National Bureau of Standards

Panel Members:
Dr. A. B. Anderson, Yale University
Dr. J. W. Gadzuk, National Bureau of Standards
Dr. D. R. Hamann, Bell Telephone Laboratories
Dr. R. D. Messmer, General Electric
Dr. J. R. Smith, General Motors

I. Opening Remarks by Chairman

The session was called to order by Chairman Watson with the announcement that "Now theorists from both the fields of chemistry and physics will have some time to claim their virtues and hide their shortcomings." The chairman explained that Prof. Grimley was not able to attend and so in his place, J. W. Gadzuk would speak on a topic about which Prof. Grimley might have spoken, had he been present. However, it was noted that Prof. Grimley might have emphasised more strongly the tenuous (at best) link between current state of the art electronic surface calculations and catalysis. This notion continued to raise its head throughout this session.

Watson suggested the following points to be dealt with, in either this session or future calculational programs. What questions can or cannot be addressed by a particular model approach? With the arrival of modern spectroscopic probes, calculations of single electron energy levels are proliferating as it is believed by many that these numbers can be compared with say observed photoemission spectra. The chairman rhetorically asked whether such energy level calculations have seen a variational principle. Do the potentials generated from the wavefunctions make sense? He stressed that good charge densities do not follow automatically from a set of well-liked one electron levels (where "well-liked" is probably defined as those agreeing with experiment).

153

With this charge, the podium was turned over to Dr. Gadzuk who spoke on the pragmatic utility of model Hamiltonian approaches for describing chemical events at surfaces, as outlined in the following report.

II. The Role of Model Hamiltonians in Chemisorption and Catalysis

J. W. Gadzuk

National Bureau of Standards

Washington, DC 20234

Great simplification in the description of the electronic structure of a coupled atom-metal system results if the chemisorbed state can be characterized in terms of some properties of the individual, uncoupled constituents and a few parameters which in principle are calculable, but in practice are usually adjusted to agree with some experimental spectroscopic data. The picture of chemisorption we envision is shown in fig. 1. The uncoupled atom and metal are shown in fig. 1a together with characteristic wavefunctions for each entity. As the atom is brought to the surface, bonding orbitals are formed between the atom and metal, which shift and broaden the originally discrete atomic level as in figs. 1b and 1c. System wavefunctions now extend throughout the atom and metal. Those states whose energies are within the resonance have disproportionately large charge densities in the vicinity of the atom.

The basic advantages of a model description of such a state of affairs are physical transparency and minimal computation. Thus the model approach is ideally suited as a testing ground for new ideas. Benchmark theoretical or spectroscopic data can easily be incorporated into the model. Such chemically desirable things as potential energy surfaces should be relatively easy to calculate.

The principle disadvantage of the approach is that it is not a "first principles theory." Thus, one has parameters available which are often treated in an arbitrary manner. (This is also true in so-called ab initio theories, although the parameters and "adjustable assumptions"

154

are hidden in more subtle ways.) In order to write down a model Hamiltonian,
one must first independently decide what is important in the chemisorption
bond. The model Hamiltonian only provides a mathematical vehicle for
displaying the physics which was already decided upon. In otherwords,
you get nothing out which was not put in. The reader is referred to
either recent review articles[1] or research papers[2] for further
details.

The most widely used model is the so-called Anderson Hamiltonian:

$$H_A = \sum_{k,\sigma} \varepsilon_k \, n_{k,\sigma} + \sum_{\sigma} \varepsilon_a n_{a\sigma}$$

$$+ \sum_{k,\sigma} (V_{ak} \, c_{a\sigma}^{\dagger} \, c_{k\sigma} + H.C.) + U \, n_{a\sigma} \, n_{a-\sigma} \; . \tag{1}$$

The various quantities in this operator, written in the occupation
number representation, are identified as follows. ε_k is the metal "band
structure" energy and $n_{k\sigma}$ is the number of electrons occupying state k.
The adatom ionization energy (suitably modified to include surface
shifts) is ε_a. The coupling term $V_{ak} = \langle a|H|k \rangle$ transfers electrons from
the adatom to metal and vice versa and is just a quantum chemical resonance
integral. The last term U is a measure of the coulomb repulsion between
electrons of opposite spin which are simultaneously on the adatom.

The field theoretic treatments provide succinct expressions for the
properties of the coupled system which quantum chemists would describe,
in the U=0 limit, by wavefunctions of the form

$$\psi_{sys} (q,k) = a(q)\psi_a(k) + \sum_{k} b(q,k)\psi_k(k) \tag{2}$$

where a and b are coefficients and q is the set of quantum numbers of
the coupled system. The main difference between the field theoretic
versus chemical approach is that k is well represented as a continuous
variable in the solid state, whereas it is taken as a discrete set of
quantum numbers in a molecule. From either point of view, a little bit
of a lot of states.ψ_{sys} are on the adatom.

The model Hamiltonian worker has well defined procedures for con-
structing various Green's functions from Eq. 1. The Green's function
associated with the electron charge on the adatom is labeled $G_{aa}(\varepsilon)$ and

$\frac{1}{\pi}$ Im $G_{aa}(\varepsilon)$ is the adatom local density of states, in its simplest form a Lorentzian shown in figs. 1b and 1c. Equivalently, the eigenvalues of Eq. 1 are given by a secular determinant in which the only off diagonal terms, $V_{a\underset{\sim}{k}}$, appear in one row and one column. These eigenvalues satisfy

$$\varepsilon - \varepsilon_a - \sum_{\underset{\sim}{k}} \frac{|V_{a\underset{\sim}{k}}|^2}{\varepsilon - \varepsilon_{\underset{\sim}{k}}} = 0 \tag{3}$$

Stated otherwise, they are the poles of the Green's function

$$G_{aa}(\varepsilon) = \left[\varepsilon - \varepsilon_a - \sum_{\underset{\sim}{k}} \frac{|V_{a\underset{\sim}{k}}|^2}{\varepsilon - \varepsilon_{\underset{\sim}{k}}} \right]^{-1} \equiv (\varepsilon - \varepsilon_a - \Lambda(\varepsilon) - i\Delta_a(\varepsilon))^{-1}$$

with the local density of states

$$\rho_{aa}(\varepsilon) = \frac{1}{\pi} \text{ Im } G_{aa}(\varepsilon) = \frac{1}{\pi} \frac{\Delta_a(\varepsilon)}{(\varepsilon - \varepsilon_a - \Lambda(\varepsilon))^2 + \Delta_a^2(\varepsilon)}. \tag{4}$$

If the level shift $\Lambda(\varepsilon)$ and width $\Delta_a(\varepsilon)$ functions are independent of $\underset{\sim}{k}$ and thus ε, then ρ_{aa} would be a Lorentzian. For most interesting cases of chemisorption, both the magnitude as well as the $\underset{\sim}{k}$ (and thus $\varepsilon(\underset{\sim}{k})$) dependence are all important in determining the electronic structure.

In the language of quantum chemistry, the local density of states on the adatom is the square of the energy resolved projection of the system wavefunction onto the atomic state. That is

$$\rho_{aa}(\varepsilon) = \sum_q |<\Psi_a|\Psi_{sys}(q)>|^2 \ \delta(\varepsilon - \varepsilon_q) = \sum_q |a(q)|^2 \delta(\varepsilon - \varepsilon_q) \tag{5}$$

which should be equivalent to Eq. 4. In both cases the orthogonality condition $<a|\underset{\sim}{k}> = 0$ has been assumed[3].

We must now make some connection between localized bonding involving both discrete and continuum states. Suppose, as shown in fig. 2a, that a hydrogenic atom tries to bridge-bond to the d_{xy} group orbitals of the substrate. The net overlap of Ψ_a with

$$\Psi_{nb} = \frac{1}{\sqrt{2}} \left[d_{xy}(1) + d_{xy}(2) \right] \tag{6a}$$

vanishes whereas the overlap and thus bonding is nonzero for the rephased
group orbital

$$\Psi_b = \frac{1}{\sqrt{2}} \left[d_{xy} (1) - d_{xy} (2) \right]. \tag{6b}$$

The resulting 3 atom molecular orbital spectrum, obtained from the
coupled group orbitals and adatom, are shown in fig. 2b. Ψ_a and Ψ_b
form bonding and antibonding orbitals whereas Ψ_{nb} remains unperturbed.
In the case of an infinite solid, the discrete energy levels merge into
a band of energies shown on the right side of fig. 2c. The Bloch eigenfunctions, written in a tightbinding representation are

$$\psi_k (\underline{r}) = \frac{1}{\sqrt{N}} \sum_n e^{i\underline{k}\cdot\underline{R}_n} \phi (\underline{r}-\underline{R}_n) \tag{7}$$

where N is the number of atoms in the solid and R_n = na (with n an
integer and a the primitive translation vector of the lattice) is the lo-
cation of atom n. For k=0, $\psi_{k=0} (\underline{r}) = \frac{1}{\sqrt{N}} \sum_n \phi (\underline{r} - \underline{R}_n)$ is just an LCAO
with the phase of all orbitals identical, as with the no-bond group
orbital. For $k = \frac{\pi}{a}$, at the zone boundary, $e^{ikR_n} = e^{i\pi n}$ and $\Psi_{k = \frac{\pi}{a}} = \frac{1}{\sqrt{N}} \sum_n e^{i\pi n} \phi(\underline{r} - \underline{R}_n)$ which is an LCAO with phases of alternate orbitals
reversed, as in the bonding group orbital.

Intuitively one should expect then that the adatom, while interacting
with the semi-infinite metal, sees the $k \approx \pi/a$ states near the bottom of
the band as bonding group orbitals and the $k \approx 0$ states near the top of
the band as non-bonding group orbitals. The resulting local density of
states might look like that shown in fig. 2c. Here a state is split
off below the band and closely resembles a localised bonding orbital.
A distribution of virtual states throughout the d-band appears with a
resonance near ε_+ which could be called an anti-bonding virtual state.
Due to a mild repulsion between ε_a and states $\underline{k} \gtrsim \underline{0}$, a localised non-
bonding state is likely to be pushed out above the band. The degree to
which discrete level cluster eignvalues (fig. 2b) resemble continuous
local densities of states (fig. 2c) determines the usefulness of a cluster
approach to this solid state problem.

If the adatom orbital couples mainly to a single group orbital of
the substrate, the hopping or resonance integral might be approximated

by

$$V_{a\underset{\sim}{k}} = \langle a|H|\underset{\sim}{k}\rangle \simeq \langle a|g\rangle \langle g|H|\underset{\sim}{k}\rangle .$$

This makes life and calculations much more manageable since $V_{a\underset{\sim}{k}}$ is now separable into a product of an atomic overlap integral $S(\underset{\sim}{R}_a) \equiv \langle a|g\rangle$ depending on the position of the adatom, but not on $\underset{\sim}{k}$ times a substrate hopping integral $F(\underset{\sim}{k},g) \simeq \langle g|H|\underset{\sim}{k}\rangle = \sum_{i,j} a_j e^{i\underset{\sim}{k}\cdot\underset{\sim}{R}_i} \langle \phi_g(\underset{\sim}{R}_j)|H|\phi(\underset{\sim}{R}_i)\rangle .$

Here the sum on j is over the centers in the group orbital with coefficients a_j. Claiming that F is a function only of $\underset{\sim}{k}$ and g, but not $\underset{\sim}{R}_a$ is an approximation since H is a function of $\underset{\sim}{R}_a$ and thus $\langle j|H|i\rangle$ does vary with R_a. However we can hope that this is a small factor since it measures the change in hopping between substrate orbitals due to the perturbation outside the solid. (This is equivalent to neglecting a $V_{kk'}$ term in the Anderson model.) With this factorization, the eigenvalues given by Eq. 3 can be written

$$\varepsilon - \varepsilon_a - |S(\underset{\sim}{R}_a)|^2 \sum_{\underset{\sim}{k}} \frac{|F(\underset{\sim}{k},g)|^2}{\varepsilon - \varepsilon(\underset{\sim}{k})} = 0 \; . \tag{8}$$

The beauty of Eq. 8 is that the sum on $\underset{\sim}{k}$ quantity, call it $Q(\varepsilon)$, can be calculated once and for all for a given substrate, group orbital, and energy. $Q(\varepsilon)$ is independent of $\underset{\sim}{R}_a$ (within our approximations). Thus the adatom Green's function becomes:

$$G_{aa}(\varepsilon) = (\varepsilon - \varepsilon_a - |S(\underset{\sim}{R}_a)|^2 Q(\varepsilon))^{-1} .$$

Kjöllerström, Scalapino, and Schrieffer[4] have given an expression for the electronic interaction energy, within the Anderson model, which with our eigenvalues is simply

$$\Delta E(\underset{\sim}{R}_a) = \frac{1}{2\pi i} \oint (\varepsilon_a + \varepsilon + |S(\underset{\sim}{R}_a)|^2 Q(\varepsilon) - 2\varepsilon$$
$$\times |S(\underset{\sim}{R}_a)|^2 Q(\varepsilon)) \times G_{aa}(\varepsilon) \, d\varepsilon - \varepsilon_a . \tag{9}$$

Equation 9 yields potential energy surfaces of an atom or molecule interacting with a surface. Such surfaces form the backbone from which reaction coordinates needed to understand kinetics, and thus catalysis, can be obtained. As it stands, the model Hamiltonian based Eq. 9 can be handled quite simply. The only function of adatom position is the overlap integral $S(R_a)$. Mulliken[5] has given formulae for overlap integrals of Slater functions on different centers. Alternatively, S can be parameterized and fitted to experimental data by setting

$$S(R_a) = S_{expt}(R_a = R_{equil}) \exp(-\beta |R_a - R_{equil}|)$$

where S_{expt} is chosen to give the correct desorption energy or to agree with spectroscopic data. The range parameter β could be determined from Mullikan's formulae. More will be heard about this procedure in Hamann's talk.

A simple example of the type of result I have in mind is the set of potential energy surfaces and reaction paths shown in fig. 3. Deuss and van der Avoird[6] considered the problem of dissociative chemisorption of H_2 on transition metal surfaces: The model they considered is the 4 atom cluster, also shown in fig. 3. In this particular calculation they calculated the interaction energy of the broadside H_2 molecule with the $3d_{z^2}$ Ni group orbital as a function of molecule surface and H-H separation. Contours of constant interaction energy are shown in fig. 3 and the dotted path is the reaction path. From this figure it is seen that if the H_2-Ni interaction is solely through a single $3d_{z^2}$ group orbital, then the H_2 will dissociate with no activation barrier to overcome. As cautioned by Deuss and van der Avoird, both this model of chemisorption and the theory used are overly simplified so the results are not to be compared with real experiments. Nonetheless, this type of calculation appears to be almost feasible with more realistic models, especially using a model Hamiltonian procedure such as that one sketched out in these notes. When this stage of development soon arrives, we should be in a much better position to really know why the electron factor in catalysis is a factor.

References

[1]   Some current review articles which stress model Hamiltonians are:
       T. B. Grimley, Adv. in Surface and Membrane Sci. 9, 71 (1975);
       J. W. Gadzuk, in "Surface Physics of Materials," Ed. J. M.
       Blakely (Academic, N.Y., 1975); R. Gomer, Solid State Phys. 30,
       94 (1975).

[2]   Examples of model Hamiltonian research papers include: P. W.
       Anderson, Phys. Rev. 124, 41 (1961); D. M. Newns, Phys. Rev.
       178, 1123 (1969); J. W. Gadzuk, J. K. Hartman, and T. N. Rhodin,
       Phys. Rev. B4, 241 (1971); G. Doyen and G. Ertl, Surface Sci. 43, 197
       (1974); W. Brenig and K. Schönhammer, Z. Phys. 267, 201 (1974); R. H.
       Paulson and J. R. Schrieffer, Surface Sci. 48, 329 (1975).

[3]   A. Madhukar, Phys. Rev. B 8, 4458 (1973); A. Bagchi and M. H.
       Cohen, Phys. Rev B 9, 4103 (1974).

[4]   B. Kjöllerström, D. J. Scalapino, and J. R. Schrieffer, Phys. Rev.
       148, 665 (1966).

[5]   R. S. Mulliken, C. A. Rieke, D. Orloff, and H. Orloff, J.Chem. Phys.
       17, 1248 (1949).

[6]   H. Deuss and A. van der Avoird, Phys. Rev. B 8, 2441 (1973).

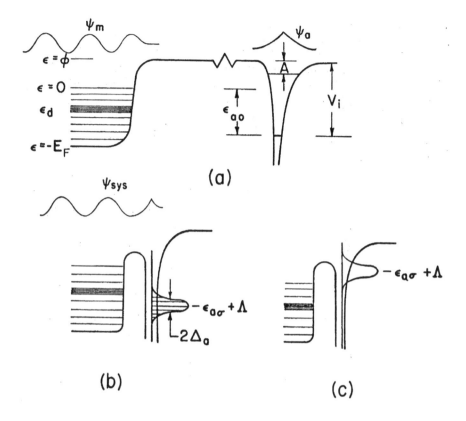

Fig. 1.   (a)   Schematic potential and energy level diagram for non-
interacting atom and metal.   The occupied portion of the
conduction band lies within the range $-E_F \leq \varepsilon \leq 0$.   A
narrow d-band is centered at $\varepsilon = \varepsilon_d$.
(b)   Adsorption for which the broadened atomic virtual
state lies below the Fermi level and is thus totally
unoccupied.
(c)   Ionic adsorption for which the broadened valence level
lies above the Fermi level and is thus almost totally
unoccupied.

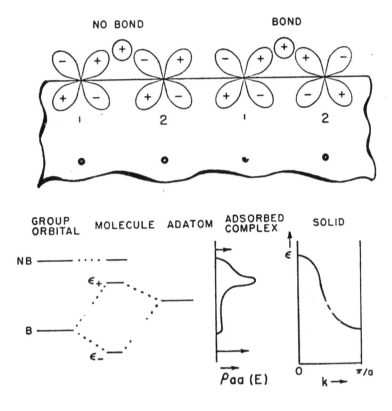

Fig. 2. (a) Adatom with a valence s orbital together with nonbonding
bridge d_{xy} group orbital (left) and bonding group orbital
(right).

(b) Molecular energy level diagram for the orbitals shown
in (a).

(c) Local density of states formed on the adatom when it
bonds to a surface through the group orbitals shown in (a).
Here the discrete bonding and non-bonding group orbital
energies are replaced by the energy band continuum.

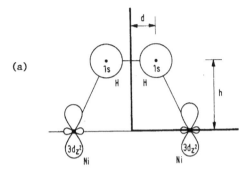

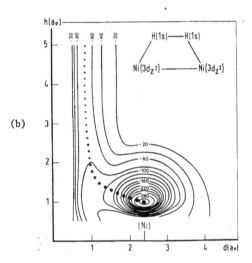

Fig. 3. (a) Model for H_2 interaction with Ni surface through the Ni d_{z^2} orbitals.

(b) Potential energy surfaces (Kcal/mole) from the model in a. The dotted line shows the most favorable reaction path followed by the H_2 molecule. The size of the dots is a measure for the value of the interaction energy.

III. Discussion Following First Paper

Dr. Messmer reiterated and added to the points brought up in the previous talk concerning the similarity in physical content, if not mathematical appearance, between the P. W. Anderson model Hamiltonian held dear by many solid state surface theorists, and simple (not extended) Hückel theory appearing in 30 year old quantum chemical papers by such people as Coulson. A translation dictionary was presented in which Hückel concepts were expressed in terms of "modern" Green's function language. For instance, the imaginary part of a Green's function, often called the local density of states, is none other than the Hückel orbital coefficients (modulus squared) in the limit of a molecule with an infinite number of centers. In addition, Hückel theorists discuss and calculate such quantities as bond order, atom polarizability, and bond polarizability which Messmer suggests is still virgin territory to the Green's function users. It was further pointed out that when the discrete levels obtained in a quantum chemical calculation on a finite molecule are Gaussian broadened, then the structure in the resulting density of states is compellingly similar to that obtained from a full band calculation, even for clusters containing only 27 (3x3x3) atoms. [Ed. note: A word of caution though; the relative peak heights and widths may be reasonable but the absolute values for the energies are often way off.]

Prof. A. B. Kunz (Univ. of Illinois) suggested that Schrieffer and Soven (Physics Today, April, 1975) had contrasted the band and broadened cluster density of states in a way which pointed out significant differences.

Dr. Messmer replied that the general shapes did agree, although edge effects in clusters could cause some discrepancies. [Ed. note: Since all but one atom in a 27 atom cluster is either at an edge or has a nearest neighbor edge atom, the resolution of this question with existing cluster calculations cannot really be achieved.]

Dr. D. R. Hamann mentioned a similar controversy related to cluster versus semi-infinite calculations on Si surface states. Calculations due to Batra and Ciraci on 14 atom Si clusters (in which the dangling bonds that would be connected to other Si atoms in a semi-infinite solid are saturated with H atoms), give surface states which differ from those obtained with the procedure of Appelbaum and Hamann, to be discussed shortly. Spatial relaxation of surface atoms give rise to new bands of surface states which appear quite differently in cluster or continuum models. Hamann does agree that clusters, when treated with care, could be fine for describing localized bonding.

Dr. Messmer concurred with Hamann and added that if one considers questions which depend on energies comparable with the width of the broadening function (or level spacing) then the cluster method cannot be expected to provide numbers which can be meaningfully compared with those generated from a continuum model.

Dr. C. B. Duke (Xerox) observed that a "Stradivarius in the hands of a village fiddler is still a fiddle." [Ed. note: It has been suggested to us by Dr. A. Melmed that a corollary to this exists: The average village audience is incapable of distinguishing between the sounds of a Stradivarius and a common fiddle, when played by a persuasive musician.] Thus with the many different kinds of cluster techniques available, those "in the hands of a skilled man are fine but in the hands of an amateur, can lead to lots of mistakes." In otherwords, you better understand your problem or "know" the answer. At this point Chairman Watson intervened with the admonition that clusters were to come later, and to keep to the schedule the remaining four presentations will be given without major interruption.

IV.   Theory For Chemisorption And Catalysis

by

Alfred B. Anderson

Chemistry Department, Yale University

New Haven, Connecticut   06520

Chemistry, by virtue of its regularities, is, at certain theoretical levels, not a difficult or perverse field of study. The regularities
in structures, binding energies, vibrational force constants and electronic energy levels in molecules have lent themselves to simple interpretations. A conceptually and computationally simple theoretical
procedure has been newly developed to deal with these properties. In it
rigid atoms are superimposed in molecular geometric configurations and
repulsive two-body forces are calculated from the charge densities
according to the Hellmann-Feynman force theorem. Charge redistributions
yield attractive energy components which might be calculated using the
Hellmann-Feynman theorem, but which are conveniently gotten as approximate one electron orbital energies. The method has been useful in
understanding small molecules and promises to be useful for large systems
because of its simplicity and low cost.

Chemical problems of catalysis on metal surfaces are complicated
and demand that theory and experiment join forces to establish rapid
progress. Three theoretical case studies are presented here. The first
is a simple orbital analysis of the catalysis of 1,3 sigmatropic shifts
by transition metal atoms, clusters and surfaces. The transition state
is stabilized through a bonding stabilization of a filled hydrocarbon
orbital with metal d orbitals. The second shows the energy levels
representing the bonding interactions and geometric distortions accompanying chemisorption of O and CO on an iron surface, compared with
experimental photoemission spectra by T. Rhodin and C. Brucker. The
third is the dissociative chemisorption of acetylene, HC ≡ CH on iron
surfaces yielding two CH groups bonding perpendicular to the surface, H

166

ends up. The calculations predict such a reaction and recent photo-emission experiments by C. Brucker and T. Rhodin show the CH σ level growing in time as acetylene dissociates on iron. Without the calculation, interpretation of the experimental spectrum is difficult. For such a reaction, with manifestly strong and localized interactions, small clusters of atoms representing a surface are adequate, but for considerations of adsordate-adsorbate interactions and coverage-dependent phenomena, larger clusters will be required.

Current References to Theory and Applications

[1] Derivation of the Extended Hückel Method with Corrections: One Electron Molecular Orbital Method for Energy Level and Structure Determinations, A. B. Anderson, J. Chem. Phys. 62, 1187 (1975).

[2] Vibrational Potentials and Structures in Molecular and Solid Carbon, Silicon, Germanium and Tin, A. B. Anderson, J. Chem. Phys. 63, 4330 (1975).

[3] Transition Metal Catalysis of Olefin Isomerizations, A. B. Anderson Chem. Phys. Letters 35, 498 (1975).

[4] Molecular Orbitals and Bonding in Ar_2, Kr_2, ArKr, $(Cl_2)_2$, ArHCl and Solid Chlorine, A. B. Anderson, J. Chem. Phys. 00, 0000 (1976).

[5] Structures, Binding Energies and Charge Distributions for Two to Six Atom Ti, Cr, Fe and Ni Clusters and their Relationship to Nucleation and Cluster Catalysis, A. B. Anderson, to be published.

[6] Interaction of Hydrogen, Carbon, Ethylene, Acetylene and Alkyl Fragments with Iron Surfaces: Catalytic Hydrogenation, Dehydrogenation, Carbon Bond Breakage and Hydrogen Embrittlement, A. B. Anderson, to be published.

167

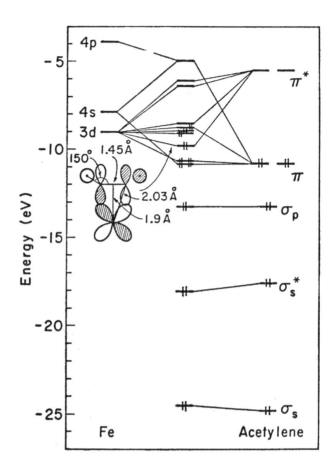

Figure 1. Molecular orbital energy level diagram for an Fe atom,
Acetylene, and acetylene bonded as shown to Fe.

168

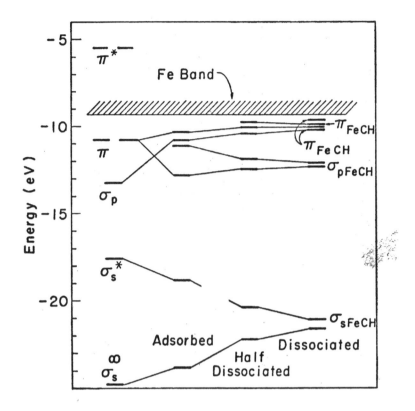

Figure 2. Molecular orbital energy level diagram for acetylene
dissociating on two Fe atoms spaced 1.866 Å. The Adsorbed
position corresponds to a C-C bond length of 1.7 Å and the
Half Dissociated position corresponds to 2.3 Å.

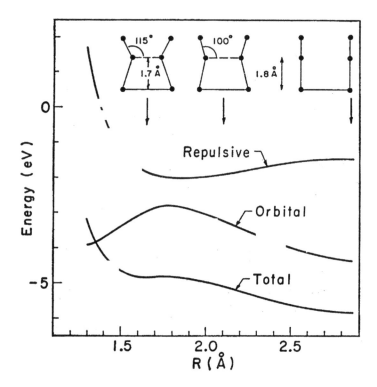

Figure 3. Total energy and components for acetylene dissociating on two Fe atoms 2.866 Å apart.

V.   Cluster Techniques Discussed by R. P. Messmer

The article reproduced as Appendix A discusses the merits of the
$X\alpha$ - scattered wave techniques, pioneered by Messmer and Johnson, in
contrast with other standard methods.   It is a good example of the type
of work presented by Dr. Messmer in his talk.

Another interesting example which was presented was a small Li
cluster, particularly since Dr. Smith showed results for the same clusters
using an alternative calculational scheme.   The following is excerpted
from an article by Messmer and K. H. Johnson, published in J. Vac. Sci.
Technol. 11, 236 (1974).

As a prelude to investigating the most catalytically important
systems, we have tested the SCF-$X\alpha$-SW computational procedure on several
simple prototype metal clusters.   For example, the simplest metal cluster
is the Li_2 diatomic molecule.   Even the most elaborate HF-SCF-LCAO
method does not yield a proper binding curve for such molecules.   In
contrast, an SCF-$X\alpha$-SW calculation (requiring only a small fraction of
the computer time expended in the HF-SCF-LCAO calculation) leads to a
total energy, equilibrium internuclear distance, and separated-atom
limit for Li_2 in relatively good agreement with experiment.

To investigate the relative stabilities of larger aggregates of Li
atoms, calculations have been carried out on clusters such as tetrahedral Li_4, square-planar Li_4, simple cubic Li_8, body-centered cubic
Li_9, cubo-octahedral Li_{13}, and icosahedral Li_{13}.

It was found that the Li_8 cluster is considerably more stable
energetically than eight separate Li atoms or four Li_2 molecules.   It is
also interesting to observe that the equilibrium Li-Li internuclear
distance is much closer to the internuclear distance in bulk crystalline
lithium than it is to the Li_2 bond length.   No experimental value for
the bond length in Li_8 is available, although there is mass-spectroscopic
evidence for the occurrence of similar alkali-metal clusters in vapor.

It has originally been shown that the SCF-$X\alpha$-SW method facilitates
the computation of molecular orbital wavefunctions and densities, thus
permitting the visual display (via computer-generated contour maps) of

electronic charge distributions and chemical bonds. This facility is particularly valuable for analyzing the fundamental nature of inter-facial (e.g., adsorbate-substrate) chemical bonds. In fig. 1, for example, we display a contour map of the valence electronic charge distribution for a Li_8 cluster, computed at the equilibrium internuclear distance and plotted in the plane of the cube face. It is especially interesting to note the significant amount of charge density located between the nuclei and directed toward the center of the cube face (indicated in fig. 1 by the contours labeled 10). The pileup of charge in the cube face is important, not only because it relates to the bonding and stability of the Li_8 cluster, but also because it may be relevant to the type of charge overlap which is essential to the reactive chemi-sorption of hydrogen on small lithium particles.

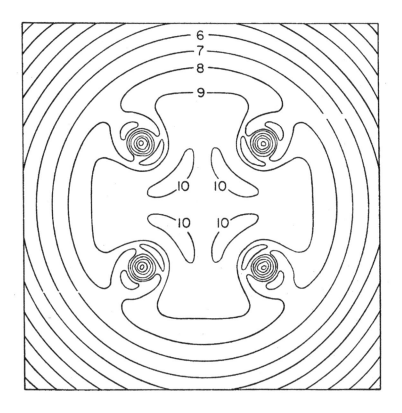

Fig. 1. Contour map of valence electronic charge density of Li_8 in the plane of a cube face, calculated by the SCF-Xα-SW method. Density of contour nearest each Li nucleus is 0.092 $\epsilon/\alpha_0 3$; density of contour 10 is 0.012 $\epsilon/\alpha_0 3$; density of contour 8 is 0.003 $\epsilon/\alpha_0 3$ (α_0=Bohr atomic radius).

Reprinted with permission of the authors and Elsevier North-Holland, Inc. from

Volume 36, number 4   CHEMICAL PHYSICS LETTERS   1 December 1975

A COMPARISON OF SCF-Xα AND EXTENDED HÜCKEL METHODS FOR METAL CLUSTERS

R.P. MESSMER*, C.W. TUCKER Jr.*

General Electric Corporate Research and Development, Schenectady, New York 12301, USA

and

K.H. JOHNSON**

*Department of Materials Science and Engineering, Massachusetts Institute of Technology,
Cambridge, Massachusetts 02139, USA*

Received 8 July 1975
Revised manuscript received 18 August 1975

A comparison of some results from Xα-scattered wave (Xα-SW) and extended Hückel (EH) calculations for metal clusters is given. It is found that small clusters of atoms (≈ 13 atoms) using the Xα-SW method reproduce many of the features of the electronic structure of the bulk metals, whereas this is not the case for the same clusters using the EH method. A more systematic approach to EH parametrizations is suggested in order to make this method a more viable approach to treating metal clusters.

Recently a number of papers have appeared in which the extended Hückel (EH) method has been used to investigate the electronic structure of transition – or noble – metal clusters [1–5] and the interaction of these clusters with adsorbates [3,5]. It had also been used previously to investigate adsorbate–substrate interactions in a non-metallic system [6]. A number of problems and shortcomings of the EH method for treating metal clusters [3] and chemisorption systems [6] have been recognized and discussed. One major problem is the proper treatment of electron transfer in a self-consistent manner. Another is the determination of the necessary parameters to treat transition metals.

In the general context of using clusters of metal atoms as a theoretical model to represent the substrate for chemisorption studies, the question arises: how many atoms are needed to give a reasonable representation of the electronic structure of a true metal? In the present letter we will compare the answers pro-

vided by calculations made to date using the EH and Xα-SW methods.

It is well known that the SCF-Xα-SW method reduces to the KKR method of band theory when applied to the perfect bulk metal [7] and that the latter has been very successfully used to describe the electronic structure of many metals [8]. Hence the Xα-SW method should provide a useful starting point for the investigation of finite metal clusters in an attempt to answer the above question. On the other hand, the extended Hückel method has not been applied properly to obtain a band structure of any metal and hence its efficacy for treating metal clusters is a priori in doubt. However, the utility of the extended Hückel method for obtaining the band structures of certain semiconductors as well as the electronic structure of finite clusters of atoms representing these semiconductors is well established [9].

The SCF-Xα-scattered wave method has been employed to investigate clusters of up to 13 atoms of Li [10], and of Cu, Ni, Pd and Pt [11]. The parameters of the Xα-SW method in its muffin-tin form consist of (i) the atomic sphere radii, (ii) the exchange parameter α and (iii) the basis set, i.e., the number of partial waves on each center. For a cluster of 13 atoms

* Research sponsored in part by the Air Force Office of
Scientific Research (AFSC) Contract F44620-72-C-0008.
** Research sponsored by the National Science Foundation,
the Petroleum Research Fund (administered by the American Chemical Society), and the Shell Foundation.

representing the nearest neighbor environment in an fcc lattice, all the atomic sphere radii are constrained by the geometry of the problem to be one-half the nearest neighbor distance of the lattice and the exchange parameter is the atomic value tabulated by Schwarz [12]. Partial waves up to $l = 2$ on the atomic centers and $l = 4$ on the outer sphere are used. This completely sets the calculations and hence there are no further parameters to affect the outcome of the calculation. This is in marked contrast to the EH method as applied to such systems, where changes in possible parametrization can yield rather different results and interpretations [1,5].

In the EH calculations of Anderson and Hoffmann (AH) charge differences on neighboring atoms of over one electron occur in some cases. No such large charge differences have been found in the $X\alpha$-SW calculations on metal clusters [11]. AH note, however, that although the charge buildups on atoms are overestimated due to the approximate nature of the EH calculations, "*the sign and relative magnitude of the charges are nevertheless useful for qualitative discussion*". Thus they clearly attribute physical or chemical significance to these charge differences; this is in contrast to other workers such as Fassaert et al. [1] who do not attribute any significance to the initial charges of the atoms in the metal cluster but only consider changes in charge distribution relative to the isolated metal cluster in discussing chemisorption.

We have repeated the calculations for Ni_9 and W_9 and reproduced the AH results. Moreover we have extended the calculations (using the AH parameters) to clusters of W_{13} and Ni_{13}. For the latter cluster we may make a direct comparison with results obtained by the $X\alpha$-SW method. In the first four rows of table 1, the EH net atomic charge results are presented for the W_9, Ni_9, W_{13}, and Ni_{13} clusters using the AH parameters. The calculated charges for Ni_{13} are of the same sign as those found from the $X\alpha$-SW calculations (see table 1); the energy levels which will be discussed below are not, however, in very good agreement. One thing which stands out immediately when comparing the first four rows of table 1 is the rather notable qualitative differences between the tungsten and nickel results. Such differences in charges according to Anderson and Hoffmann should be physically meaningful, if not in a quantitative sense nevertheless for qualitative discussion.

Table 1
EH net atomic charges for W and Ni clusters [a]

Method	Cluster	Center atom	In-plane atoms [b]	Out-of-plane atoms [c]
AH	W_9	+0.74	+0.41	−0.59
AH	W_{13}	+1.52	−0.39	+0.39
AH	Ni_9	−0.25	−0.11	+0.17
AH	Ni_{13}	−0.17	+0.01	+0.01
SZAH	Ni_9	+0.38	+0.20	−0.30
SZAH	Ni_{13}	+2.73	−0.23	−0.23
FVA	Ni_{13}	+2.54	−0.21	−0.21
SCF-$X\alpha$	Ni_{13}	−0.72	+0.06	+0.06

[a] The clusters for W and Ni have somewhat different geometries because W has a bcc structure and Ni an fcc structure.
[b] The four atoms which are in the same plane as the center atom
[c] The four (or eight) atoms which are in the plane(s) above (and below) the plane containing the center atom.

There is, however, an anomalous difference in basis functions between the AH calculations for tungsten and nickel. In the former case a single Slater function (single zeta $\equiv$ SZ) is used to represent the d-orbital whereas for nickel a double zeta (DZ) function is used. When the Ni_9 and Ni_{13} calculations are repeated using an SZ function for Ni ($\zeta = 2.0$) and keeping all other AH parameters the same, which is comparable to the AH tungsten calculations, a rather different charge distribution is obtained (see table 1). In comparing rows 3 and 4 with 5 and 6 of the table we find that the net charges change not only in magnitude but also in sign and that the SZAH results are qualitatively similar to the single zeta results of AH for tungsten. Thus we are led to the conclusion that the large differences in net charges in the AH results between Ni and W clusters are *not physically significant but reflect differences in parametrization*. This is further supported by the results of ref. [1] for an Ni_{13} cluster using the EH method, but with a different parametrization. The resultant charges are given in the seventh row of table 1 and labelled FVA.

In fig. 1 a comparison is provided for the one electron energies obtained from the SCF-$X\alpha$-SW and EH calculations. The EH calculations shown are for the AH parameters. To the left all the occupied valence levels are shown, to the right the higher occupied levels are shown on an enlarged scale. The comparison

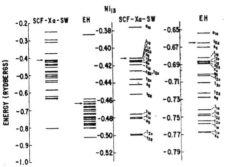

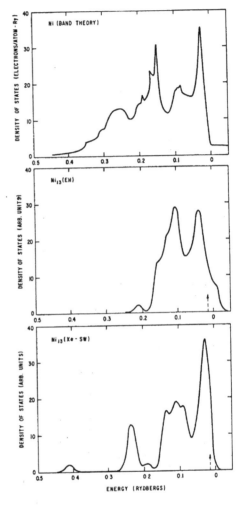

Fig. 1. A comparison of energy levels for Ni_{13} as determined by the spin-restricted SCF-Xα-SW and EH methods. The arrows indicate the calculated Fermi levels. In the EH results the highest t_{2g} level is completely occupied. In the SCF-Xα-SW results the highest t_{1g}, t_{1u}, and a_{2g} levels are nearly degenerate and have an occupancy of 8/14.

between the two calculations shows rather little agreement.

Returning to the question posed earlier of how well a small cluster represents the electronic structure of a metal, it is convenient to present the results of fig. 1 in a somewhat different form in order to compare with the density of states (DOS) of bulk nickel. Namely, we calculate the density of states for the Ni_{13} cluster by replacing each discrete eigenvalue (see fig. 1) by a gaussian with a width of 0.01 rydberg, weighted by the degeneracy of the orbital. The result is shown in fig. 2 where it is compared to the DOS results from the SCF-LCAO-Xα band structure of Callaway and Wang for bulk Ni [13]. As the extended Hückel method is a spin-restricted procedure we compare here only the Xα-spin restricted results and likewise use only the majority-spin DOS results of Callaway and Wang rather than their total spin-unrestricted results. In a spin-restricted calculation there would be the same number of majority and minority spins and hence no shift in the minority spin DOS with respect to the majority spin DOS. A comparison such as presented in fig. 2 represents one very useful and graphic criterion for assessing the similarity in the electronic structures of clusters of atoms to that of the bulk metal.

It is clear from fig. 2 that the results of EH and Xα-SW for Ni_{13} are quite different and that the Xα-SW

Fig. 2. A comparison of the density of states of Ni as determined from: (i) a bulk band structure calculation [13], (ii) a 13-atom Ni cluster using the extended Hückel method and (iii) a 13-atom Ni cluster using the SCF-Xα-SW method. The energy scales for the clusters have been shifted so as to line up the Fermi levels which are indicated by arrows.

results have many features in common with the bulk DOS, whereas this is not the case for the EH results.

$X\alpha$-SW calculations have also been performed for Cu_{13}, Pd_{13}, and Pt_{13} and will be discussed in detail elsewhere [11]. It is of interest here, however, to briefly compare some results of these calculations for Cu_{13} and Pd_{13} with recent EH calculations for Cu and Pd clusters by Baetzold and Mack [4]. Experimentally it is known that the d-band width increases through the series Cu, Ni, Pd, Pt — this trend is reproduced by the $X\alpha$-SW calculations for 13-atom clusters [11]. The results for even larger clusters, i.e., 19 atoms by Baetzold and Mack (BM) using the EH method give results for Cu and Pd which are inconsistent with experiment and with the Ni results using the AH parameters. BM calculate d-band widths for Cu_{19} and Pd_{19} of ≈ 0.2 eV and ≈ 0.3 eV, respectively (cf. fig. 9 of ref. [4]), whereas the experimental d-band widths as well as those calculated by the $X\alpha$-SW method are of the order of several eV. BM state: "In most properties, the clusters in the size range reported here are different from the bulk metals". Hence this seems to represent the answer to the question posed above — as provided by extended Hückel calculations. This conflicts, however, with the answer which emerges from $X\alpha$-SW calculations carried out thus far. These latter calculations suggest that much of the bulk band width and DOS structure can be obtained with a cluster of approximately a dozen atoms.

We suggest that this discrepancy arises from the EH calculations due to the current arbitrariness in parametrizations [1,5] used for transition metals in this method. A much more systematic approach to the problem would be to obtain parameters by matching the occupied bands as obtained from an EH band structure calculation of the metal with those obtained from more rigorous band calculations, in much the same spirit as previously used for semiconductors [9]. These parameters would then provide a reasonable starting point for calculating clusters of metal atoms. Alternatively, EH parameters might be chosen by matching to the results of $X\alpha$-SW calculations on clusters.

The importance of having a reasonable description

of the electronic structure of a metal cluster, before using this cluster to study the chemisorption of molecules cannot be over-emphasized. The inadequacy of the AH parametrization (as seen in table 1 and fig. 2) must be a strong contributing factor, along with the effects of self-consistency in charge transfer, to the fact that the Anderson—Hoffmann explanation of the photoemission results for CO chemisorbed on Ni is inconsistent with the most recent and definitive experimental data [14].

References

[1] D.J.M. Fassaert, H. Verbeek and A. van der Avoird, Surface Sci. 29 (1972) 501.
[2] R.C. Baetzold, J. Catal. 29 (1973) 129.
[3] L.W. Anders, R.S. Hansen and L.S. Bartell, J. Chem. Phys. 59 (1973) 5277.
[4] R.C. Baetzold and R.E. Mack, J. Chem. Phys. 62 (1975) 1513.
[5] A.B. Anderson and R. Hoffmann, J. Chem. Phys. 61 (1974) 4545.
[6] A.J. Bennett, B. McCarroll and R.P. Messmer, Surface Sci. 24 (1971) 191.
[7] K.H. Johnson and F.C. Smith Jr., in: Computational methods in band theory, eds. P.M. Marcus, J.F. Janak and A.R. Williams (Plenum Press, New York, 1971) p. 377.
[8] B. Segall and F.S. Ham, in: Methods in computational physics, Vol. 8, eds. B. Alder, S. Fernbach and M. Rotenberg (Academic Press, New York, 1968) ch. 7.
[9] R.P. Messmer, Chem. Phys. Letters 11 (1971) 589; G.D. Watkins and R.P. Messmer, in: Computational methods for large molecules and localized states in solids, eds. F. Herman, A.D. McLean and R.K. Nesbet (Plenum Press, New York, 1973) p. 133; R.P. Messmer and G.D. Watkins, Phys. Rev. B7 (1973) 2568; C. Weigel, R.P. Messmer and J.W. Corbett, Phys. Stat. Sol. 57b (1973) 455.
[10] J.G. Fripiat, K.H. Chow, M. Boudart, J.B. Diamond and K.H. Johnson, J.Mol. Catal., submitted for publication.
[11] R.P. Messmer, S.K. Knudson, K.H. Johnson, J.B. Diamond and C.Y. Yang, to be published.
[12] K. Schwarz, Phys. Rev. B5 (1972) 2466.
[13] J. Callaway and C.S. Wang, Phys. Rev. B7 (1973) 1096.
[14] T. Gustafsson, E.W. Plummer, D.E. Eastman and J.L. Freeouf, Bull. Am. Phys. Soc. 20 (1975) 304.

VI. Essential Points Presented by J. R. Smith

The approaches to the surface electronic structure problem discussed in this presentation are based on models that are simple enough
to be solved self-consistently from first principles, i.e., with no
adjustable parameters. The jellium model is rather accurate for work
function calculations. A comparison between calculated and measured
values is shown in fig. 1. It fails for all but a few alkalies in
surface energy calculations, however, due to the energetic consequencies
of replacing the discrete periodic lattice by a uniform smeared out
charge. In an attempt to correct for this, a pseudopotential perturbation
calculation has been performed which yields reasonably accurate surface
energies and adhesive energy profiles. Bulk ion core pseudopotentials
were used. The question of the general validity of bulk pseudopotentials
in surface calculations must still be answered[1]. A linear response
function has been computed for the jellium model and applied to hydrogen
chemisorption[2]. The interaction energy versus distance from the surface
is shown in fig. 2. It should be possible to establish a connection
between the linear response function and the calculation of potential
surfaces. To study catalysis on transition metals, one must go beyond a
(crystalline) jellium model. The generalized Wannier function (GWF)
method is a promising approach. The efficiency and accuracy of the GWF
method has been tested on a one-dimensional calculation[3]. The first
three-dimensional calculation is done on a lithium eight atom particle.
Such a particle was treated earlier[4]. Our cohesive energy is $\sim$12%
larger. A plausible explanation is that the larger value is due to the
removal of the muffin tin constraint. Our charge contours (shown in Fig.
3) in the surface (chemisorptive) region of the particle reflect much
more strongly the cubic symmetry of the particle rather than a spherical
symmetry. The method is currently being applied to adsorbate covered
transition metal surfaces.

Recommended entries to the literature: *Interactions of Metal
Surfaces*, Vol. 4, Topics in Applied Physics, edited by R. Gomer (Springer-
Verlag, NY, 1975). *Surface Physics of Crystalline Solids*, edited by J.
M. Blakeley (Academic, NY, 1975).

[Ed. note: As Smith emphasized, the non-spherical character of the
valence electron density around a Li site is predicted to be quite
different in his calculations compared with those of Messmer and
Johnson. In fact, viewing one or the other plot, one would make rather
different statements concerning the character of the valence states
which are available for bonding with adatoms. Band theory has had
difficulty in predicting aspherical charge character in the bulk of
solids as well. For example, Marvin Cohen and coworkers have found it
necessary to introduce a non-local pseudopotential in order to reproduce
the aspherical bonding density inferred experimentally for Si; standard
local pseudopotentials did not suffice.]

---

[1] N.D. Lang, Solid State Phys. $\underline{28}$, 225 (1973).

[2] S. C. Ying, J. R. Smith, and W. Kohn, Phys. Rev. B $\underline{11}$, 1483 (1975).

[3] J. R. Smith and J. G. Gay, Phys. Rev. B $\underline{12}$, 4238 (1975).

[4] K. H. Johnson and R. P. Messmer, J. Va. Sci. Technol. $\underline{9}$, 561 (1974).

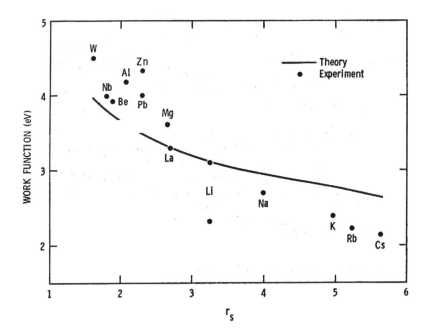

Fig. 1. Workfunction versus r_s, the nondimensional inter-electron
spacing from jellium calculations and from experiment.

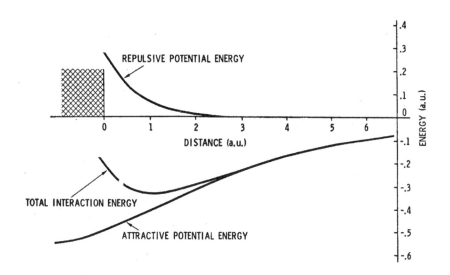

Fig. 2. Proton-metal interaction energy versus separation distance. The nuclei of the surface plane of the metal are located at $-d/2$, where d is the distance between planes parallel to the surface.

Fig. 3. Contours of constant valence charge density in the cube face of
a Li_8 cluster. The nonband charge contours are in units of
$electrons/nm^3$. Note that in the interstitial regions, the
charge density peaks in the near-neighbor directions. In the
region outside the atoms, note that the contours are more
square than circular. Outside contour 20, the value of
successive contours has a ratio of 2.

182

VII. A New Role For Theory In Surface Science

by

D. R. Hamann
Bell Laboratories
Murray Hill, New Jersey 07974

Several recently developed theoretical techniques are presently
yielding detailed information about the structural, chemical, and
spectroscopic properties of semiconductor surfaces. The chemisorption of H on the Si(111) surface and the reconstruction of the
Si(100) surface are discussed, emphasizing the interplay of theory
and experiment.

Great advances have been made in the last few years in the application of theoretical techniques to provide detailed information about
the electronic structure of solid surfaces and their chemical activity.
Progress has been most rapid in the field of semiconductor surfaces,
which has today reached the point at which meaningful comparisons and
cross-stimulation between experimental and theoretical studies are
regularly taking place. I will discuss two examples from this field in
a quasi-historical fashion--the chemisorption of H on the Si(111) surface and the reconstruction of the Si(100) surface--after briefly describing the most productive current theoretical approaches. These have
as a common thread a characteristic that I feel is extremely important:
they employ realistic models of specific systems, and methods which have
been demonstrated to produce accurate electronic structure for these
solids in the bulk.

The first approach, developed by Joel A. Appelbaum and myself,
employs a semi-infinite geometry[1]. Its basic assumption, which has
been verified by explicit calculation, is that the disturbance produced
by a surface is "healed" within several atomic layers[2]. A fully self-
consistent quantum mechanical calculation is performed, representing the
ion cores by model potentials, treating the Hartree potential of the

183

valence electrons exactly and approximating the exchange and correlation potential by a local function of the electron density. No artificial discontinuities or "muffin tins" are introduced. An approach employing an essentially similar procedure to model the physical problem, but using a slab geometry and different calculational techniques, has recently been applied to a variety of problems by Cohen and coworkers[3].

The empirical tight-binding approach has recently been applied to several surface problems by Pandey and Phillips using a slab geometry[4]. While not a new method, these workers showed for the first time that a very good fit to the entire valence band spectrum could be obtained using a small number of parameters and a sophisticated fitting procedure. They also demonstrated that a simple overlap scaling of matrix elements could represent the effects of changes in bond lengths at the surface, and that matrix elements fitted to molecular levels could be transferred to surface situations to describe chemisorption.

The story of H chemisorption on Si(111) begins, logically at least, with the ultraviolet photoemission measurement of the surface density of states by Rowe and Ibach[5]. They showed that atomic H readily adsorbs on this surface, and adds a broad peak to the density of states centered 5 to 6 eV below the valence band maximum. The geometry of the chemisorptive bond seemed simple in this case--a single bond is broken at each surface atom when the surface is formed, and each broken bond can be saturated by a single H. Adapting this geometry, Appelbaum and I calculated the electronic structure of the surface with an ordered monolayer of H[2]. The self-consistent potential found in this calculation is shown in fig. 1. By varying the normal spacing of the H layer, we found the bond length at which the forces on the layer went to zero, and the bond stretching frequency. The length was in excellent agreement with the empirical chemical value, and the force constant in agreement with infrared measurements[6]. The density of states spectrum, shown in fig. 2, presented a problem, however. It showed a distinct two peak structure, unlike the single peak in Rowe and Ibach's data[5].

Publication of these results stimulated Hagstrum and Sakurai to attempt an independent measurement of the surface density of states using ion neutralization spectroscopy. They simultaneously performed a

UV photoemission measurement, however, and found, to everyone's surprise, a two-peaked spectrum as predicted by the theory[7]. The ion neutralization spectrum proved difficult to interpret, however. Returning to the theoretical results, Appelbaum and I showed that this spectrum could be fit if one assumed that the neutralizing electron tunneled 2 Å out into the vacuum, but that the Auger-emitted electron originated within the surface layer[8]. An apparent anomalously large shift in the high energy threshold of the emitted electrons in going from clean to H covered Si was also explained by these calculations, and shown to be simply a large reduction in the amplitude of the spectrum over a 2 eV range, and not an actual threshold shift.

The next chapter in the Si-H story came when Hagstrum and Sakurai found that a Si surface prepared by quenching from high temperatures could adsorb a great deal of additional hydrogen beyond that required to saturate the amplitude of the previously mentioned two peak structure. This structure disappeared, and two new large peaks appeared at considerably lower energies[9]. Pandey showed that these puzzling results could be explained by assuming that the surface Si layer was eroded away, and that three H atoms bonded to the three available bonds of what was originally the second Si layer. A spectrum calculated using this geometry produced peaks of the correct position and shape[9].

The latest chapter in this story, a joint experimental effort by Hagstrum and Sakurai and theoretical effort by Appelbaum and me, has, in a sense, closed the circle. A partial H monolayer displays distinctly different spectra depending on whether it is ordered or disordered[10], and this then-unappreciated effect explains the difference between the initial photoemission spectrum of Rowe and Ibach[5] and the later results of Hagstrum and Sakurai[9].

The second topic which I will discuss concerns the Si(100) surface, which has long been known to occur only in a reconstructed form with a doubled translational periodicity. Two bonds are broken for each atom on this surface, so the ideal geometry is clearly not an optimum one. Two models for the atomic geometry of the reconstructed surface were proposed quite a while ago by Schlier and Farnsworth[11]. In one model,

185

pairs of surface atoms move together to rebond one of the broken bonds on each, while bending but not stretching their bonds to the second layer. In the second, half of the surface atoms are removed, so that those remaining can saturate all the broken bonds by forming double bonds. The pairing model has recently been supported by Levine in the course of explaining the extremely low work function of Si(100) with coadsorbed Cs and O[12]. The vacancy model has recently been re-introduced by Phillips who argues that it is favored by thermochemical data[13].

Appelbaum, Baraff and I began to study this surface by calculating the electronic structure of the ideal geometry[14]. Two partially occupied bands of surface states were found. While such a "metallic" surface should generally be unstable, no instability corresponding to the 2x1 reconstruction mode appeared when the dielectric response of these bands was calculated.

We proceeded to calculate the electronic structure of the recon-structed surface for both the pairing and vacancy models[15]. The general nature of the states found for the latter destroyed the hypothesis that double bonds could form. Only one additional bond resonating between the two bond directions of the surface atom to the second layer occurred, and the remaining electron pair occupied two partially filled surface state bands. The pairing model, on the other hand, did better than anticipated in its bonding. The bonds bent towards each other joined to look like a normal bulk Si bond, as may be seen from the valence charge density shown in fig. 3. The other two broken bonds of the pair formed two nearly split surface state bands, and contributed an additional nearly saturated pi-like bond. A comparison of the surface region density of states for each model with UV photoemission data taken by Rowe[16] is shown in fig. 4. While matrix element effects wipe out the photoemission from the low-lying s-like bands, the structure from the higher p-like bands clearly is better fit by the pairing model.

Although seemingly complete, the results described had an annoying loose end. The surface was still metallic because the bands of bonding and antibonding pi-like surface states overlapped slightly. With the increased bonding from this band, it seemed plausible to expect a pair

186

bond length somewhat shorter than the assumed single bond length. Another calculation was carried out for a shortened bond, and while the overlap decreased, the bands did not separate[17].

At this time, we learned of new LEED data by Webb[18], which showed that very clean and carefully annealed surfaces display an additional very weak set of spots indicating a fourth order reconstruction super-imposed on the 2x1 structure. We calculated the dielectric response of our overlapping bands of surface states for the pairing model, and found a strong tendency to instability for a deformation with just the needed periodicity to explain the additional spots[19]. The instability persisted for both choices of pair bond length. This result strongly suggests that a charge density wave state such as found in layered transition metal compounds[20] exists on this surface and is responsible for the higher-order reconstruction.

The most fascinating aspect of the Si(100) story is that its re-construction apparently involves two different members of a yet to be enumerated list of reconstruction mechansims. A "first order" effect involving a major repopulation of states takes it from the grossly unstable ideal structure to the paired 2x1 geometry. A "second order" instability effect then takes over to produce additional small atom displacements with a longer periodicity. This deformation presumably wipes out all the remaining metallic character of the surface, and leaves a stable electronic structure.

The two examples discussed indicate the kinds of understanding of the detailed physics and chemistry of surfaces that can be achieved today. The interplay of chemical effects, geometry, and spectra can be untangled through the use of adequate theoretical tools and an active give-and-take between theory and experiment. There is every reason to be confident that continuing efforts will explain many additional effects for semiconductor surfaces, and that a similar approach to the study of transition metal surfaces will meet with success in the near future.

References

[1]   J. A. Appelbaum and D. R. Hamann, Phys. Rev. B6, 2166 (1972); Phys. Rev. Lett. 31, 106 (1973); ibid. 32, 225 (1974); in Proc. of the Twelfth International Conference on the Physics of Semiconductors, ed. by M. H. Pilkuhn (B.G. Teubner, Stuttgart, West Germany, 1974), p. 675.

[2]   J. A. Appelbaum and D. R. Hamann, Phys. Rev. Lett. 34, 806 (1975).

[3]   M. Schluter, et al., Phys. Rev. Lett. 34, 1385 (1975); S. G. Louie and M. L. Cohen, ibid. 35, 866 (1975).

[4]   K. C. Pandey and J. C. Phillips, Phys. Rev. Lett. 32, 1433 (1974); ibid. 34, 1450 (1975); and to be published.

[5]   H. Ibach and J. E. Rowe, Surf. Sci. 43, 481 (1974).

[6]   G. E. Becker and G. W. Gobeli, J. Chem. Phys. 35, 458 (1961).

[7]   T. Sakurai and H. D. Hagstrum, Phys. Rev. B12, (1975).

[8]   J. A. Appelbaum and D. R. Hamann, Phys. Rev., to be published.

[9]   K. C. Pandey, T. Sakurai and H. D. Hagstrum, to be published.

[10]   J. A. Appelbaum, H. D. Hagstrum, D. R. Hamann, and T. Sakurai, to be published.

[11]   R. E. Schlier and H. Farnsworth, in Semiconductor Surface Physics, ed. by R. H. Kingston (University of Pennsylvania Press, Philadelphia, 1957); J. Chem. Phys. 30, 917 (1959).

[12]   J. Levine, Surf. Sci. 34, 90 (1973).

[13]   J. C. Phillips, Surf. Sci. 40, 459 (1973).

[14]   J. A. Appelbaum, G. A. Baraff and D. R. Hamann, Phys. Rev. B11, 3822 (1975) and to be published.

[15]   J. A. Appelbaum, G. A. Baraff and D. R. Hamann, Phys. Rev. Lett. 35, 729 (1975).

[16]   J. E. Rowe, Phys. Lett. 46A, 400 (1974).

[17]   J. A. Appelbaum, G. A. Baraff and D. R. Hamann, to be published.

[18]   M. B. Webb, private communication.

[19]   J. A. Appelbaum, G. A. Baraff and D. R. Hamann, to be published.

[20]   J. A. Wilson, F. J. DiSalvo and S. Mahajan, Phys. Rev. Lett. 32, 882 (1974); D. E. Moncton, J. D. Axe and F. J. DiSalvo, Phys. Rev. Lett. 34, 734 (1975).

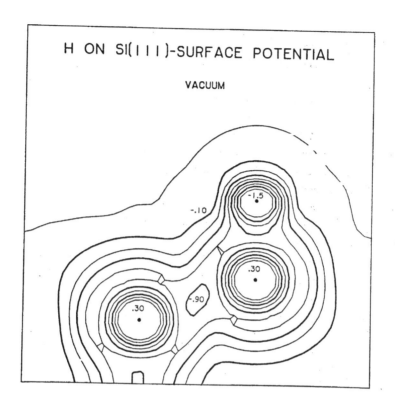

Fig. 1.　Contour plot of the self-consistent potential of H chemisorbed
on Si(111). The plane of the plot is normal to the surface, and
the dots indicate the positions of the H, first-layer Si, and
second-layer Si atoms. The contours are at 0.2 Hartree
intervals, and the scale is such that the valence-band
maximum falls at +0.066.

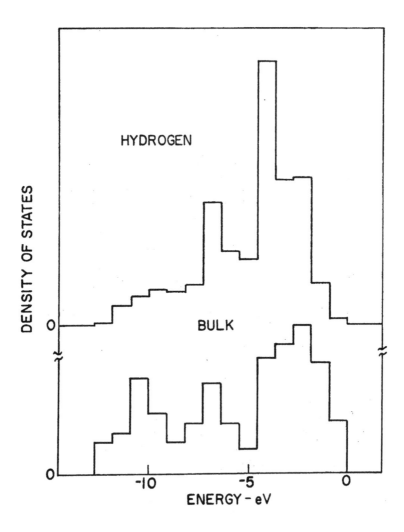

Fig. 2. Valence-band density of states for bulk Si and local density
of states on the chemisorbed H atoms. The histograms are
normalized for equal areas.

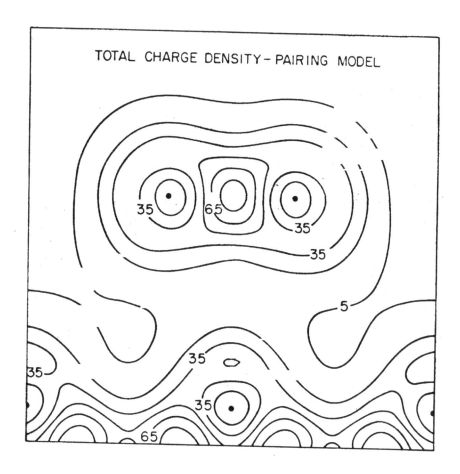

Fig. 3. Charge-density contours on a plane normal to the 2x1 reconstructed Si(100) surface passing through the paired surface atoms and fourth-layer atoms (shown by dots). Second- and third-layer atoms lie out of this plane. Density is in atomic units $x10^{-3}$.

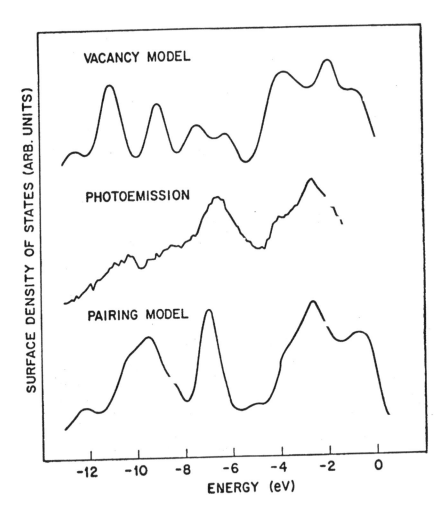

Fig. 4. Calculated surface-region density of states for the 2x1 reconstructed Si(100) surface compared to $\hbar\omega$ = 21.2 eV photoelectron-
energy distribution from Ref. 16, with estimated secondaries
subtracted.

VIII. Open Discussion Following Presentations

Prof. M. Boudart (Stanford), sensing a defensiveness on the part of
the solid state theorists, indicated that jellium calculations may not be
as far afield from catalysis as some people might think, at least if
jellium could be prepared in a dispersed state. For instance, Na dispersed on alumina is an excellent catalyst for reactions involving
carbon ion intermediates. Also, one should not belittle cluster calculations, because people like John Sinfelt deal with clusters daily in real
life. Clusters are much more important to those in catalysis than the
ideal virgin surface of the surface physics laboratory. The comments
inspired Dr. Duke to add a few words of caution related to the ubiqui-
tous adjustable parameters in most models. One must ask the question of
how these parameters are determined? A current method is to adjust the
parameters until the calculations agree with an experimental photoemission
spectrum. However, one must still know if both the spectrum and its
cluster geometry are simultaneously correct. It was suggested that
there are a broad class of models which get the spectrum right but not
the geometry. This is okay for single crystal work since you already
know the geometry. On the other hand, for catalytic systems it is important to have a model which gets both the spectra and the geometry
right.

Prof. Anderson noted that he obtained his atomic parameters from
Clementi wavefunctions and required his clusters to have the correct
diatomic equilibrium distances, force constants, and binding energies.

Prof. John Turkevich (Princeton University) offered some questions
to focus on. One is 100 years old. Does the catalyst distort the
molecule? Is the distortion of the molecule you put on the surface
different from atom to atom? Are electrons transferred more easily?
Chemisorption is secondary to catalysis. What makes a molecule break up,
what makes it active, what makes O_2, H_2, N_2, which are dead, all of a
sudden react? What is the activation process? Maybe by going to really
simple things you may reduce the whole essence of the problem. There
must be an irreducible minimum where the essence of the catalytic act,
[Ed. note: natural or unnatural] as we practice it, is retained but is

193

not that complicated that we can't swallow it. Photoelectron spectra
are very interesting for surface chemistry, for all sorts of solid state
effects, but that's not catalysis.

Dr. Duke interjected that models of catalysis are going to be
checked for a small particle by some sort of spectroscopy. The reply
from Dr. Turkevich, "yes, but spectroscopies are very indirect. It may
give you a living but it will not give us a living" summarized the gulf
(which has slowly narrowed in the past year or two) still existing
between the work of the surface scientist and the catalytic scientist.

WORKSHOP ON THE ELECTRON FACTOR IN CATALYSIS ON METALS
December 8-9, 1975

REGISTRATION LIST

Alayne A. Adams
American University
Dept. of Chemistry
Massachusetts & Nebraska Ave., NW
Washington, DC   20016

Radoslav Adzic
Institute of Electrochemistry
Karnegieva 4
Belgrade, Yugoslavia   11001

A. W. Aldag
University of Oklahoma
202 W. Boyd
Norman, OK   73069

Alfred B. Anderson
Yale University, Chemistry Dept.
225 Prospect St.
New Haven, CT   06520

Joseph A. Baglio
GTE Labs
40 Silvan Avenue
Waltham, MA   02154

Alfonso L. Baldi
Alloy Surfaces Company
100 S. Justison Street
Wilmington, DE   19899

Ian R. Bartky
National Bureau of Standards
Matls Bldg., B354
Washington, DC   20234

Lawrence H. Bennett
National Bureau of Standards
Matls Bldg., B152
Washington, DC   20234

Harold Berger
National Bureau of Standards
Matls Bldg., B254
Washington, DC   20234

George Birnbaum
National Bureau of Standards
Matls Bldg., B354
Washington, DC   20234

M. Boudart
Stanford University
Stanford, CA   94305

E. O. Box, Jr.
Phillips Petroleum Co.
334A Res. Bldg #1
Research Center
Bartlesville, OK   74003

Manfred W. Breiter
General Electric Co.
Corp. Research & Development
P.O. Box 8
Schenectady, NY   12345

Boris D. Cahan
Case Western Reserve Univ.
2074 Adelbert Road
Cleveland, OH   44106

Julio Calderon
Dept. of the Army
AMXFB-CF
Ft. Belvoir, VA   22060

Patrick J. Call
RCA Laboratories
Washington Road
Princeton, NJ   08540

Robert C. Cargill
Physical Elect. Industries, Inc.
278 Fairtree Plaza
Severna Park, MD   21146

Divna Cipris
Allied Chem. Corp.,
Columbia Road
Morristown, NJ   07960

Paul H. Citrin
Bell Laboratories
600 Mountain Avenue
Murray Hill, NJ   07974

William L. Clinton
Georgetown University
37 & 0 Streets, NW
Washington, DC   20057

Ernest M. Cohn
103 G Street, SW
Apt. B-620
Washington, DC   20024

Amos Coleman
Dept. of the Army
AMXFB-CF
Ft. Belvior, VA   22060

Bernard R. Cooper
West Virginia University
Dept. of Physics
Morgantown, WV   26506

John R. Cuthill
National Bureau of Standards
Matls Bldg., B148
Washington, DC   20234

Victor V. Damiano
Franklin Institute Res. Labs
20th & Benjamin Franklin Pkwy
Philadelphia, PA   19103

H. L. Davis
Solid State Div.
Oak Ridge National Lab.
Oak Ridge, TN   31830

Jean F. Delord
Reed College
3203 SE Woodstock Blvd.
Portland, OR   97222

Thomas W. DeWitt
National Science Foundation
1800 G Street, NW
Washington, DC   20550

Richard A. Diffenbach
ERDA
4800 Forbes Avenue
Pittsburgh, PA   14213

C. B. Duke
Xerox Corp.
800 Phillips Road, W-114
Webster, NY   14580

Henry Ehrenreich
Harvard University
Cambridge, MA   02138

T. L. Einstein
Dept. of Physics & Astronomy
University of Maryland
College Park, MD   20742

Nils E. Erickson
National Bureau of Standards
Chem Bldg. B248
Washington, DC   20234

Gerhard Ertl
Universitat Munchen
Sophienstr. 11
8 Munchen 2, W. Germany

Billy J. Evans
National Bureau of Standards
Matls Bldg., B150
Washington, DC   20234

Galen B. Fisher
National Bureau of Standards
Chem Bldg., B248
Washington, DC   20234

Robert Mark Friedman
Monsanto Company
Mail Zone Q3B
800 N. Lindbergh Blvd.
St. Louis, MO   63166

Yasuo Fukuda
Center of Materials Research
University of MD
College Park, MD   20742

J. William Gadzuk
National Bureau of Standards
Metrology Bldg., B214
Washington, DC   20234

C. D. Gelatt
Harvard University
Pierce Hall
Cambridge, MA   02138

Heinz Gerischer
Fritz-Haber-Institute der
 Max-Planck-Gesellschaft
Faradayweg 4-6
1 Berlin 33, W. Germany

William L. Gordon
Case Western Reserve Univ.
Dept. of Physics
Cleveland, OH   44106

Warren Grobman
IBM Thomas Watson Res. Ctr.
P.O. Box 218
Yorktown Heights, NY   10598

Laszle Guczi
Worcester Polytechnic Institute
Chem. Eng. Dept.
Worcester, MA   01609

Gary L. Haller
Yale University
Dept. of Eng. & Applied Science
9 Hillhouse Ave.
New Haven, CT   06520

Robert S. Hansen
Ames Laboratory, USERDA
Iowa State University
Ames, IA   50011

D. R. Hamann
Bell Labs
Murray Hill, NJ   07974

Victor E. Henrich
MIT, Lincoln Lab.
Wood Street
Lexington, MA   02155

Jan F. Herbst
National Bureau of Standards
Metrology Bldg., B214
Washington, DC   20234

John D. Hoffman
National Bureau of Standards
Matls Bldg., B364
Washington, DC   20234

Emanuel Horowitz
National Bureau of Standards
Matls Bldg., B364
Washington, DC   20234

Howard S. Jarrett
E.I. duPont de Nemours & Co.
Experimental Station
Wilmington, DE   19898

Leslie H. Jenkins
Oak Ridge National Laboratory
P.O. Box X
Oak Ridge, TN   37830

Johann A. Joebstl
Electrochemical Div/Lab 3000
AMXFB-EC, USAMERDC
Ft. Belvoir, VA   22060

Harlan B. Johnson
PPG Industries
P.O. Box 31
Rittman, OH   44203

James R. Katzer
Department of Chemical Engineering
University of Delaware
Newark, DE   19711

Richard D. Kelley
National Bureau of Standards
Chem Bldg., B248
Washington, DC   20234

Chris E. Kuyatt
National Bureau of Standards
Metrology Bldg., B212
Washington, DC   20234

197

M. G. Lagally
University of Wisconsin
1109 Eng. Res. Bldg.
Madison, WI   53706

John Lambe
Ford Motor Company
P.O. Box 2053
Dearborn, MI   48121

Lennart Larsson
University of Lund, Chem. Ctr.
P.O.B. 740
S-220 07 Lund 7, Sweden

Richard V. Lawrence
Dept. of the Army
AMXFB-CF
Ft. Belvoir, VA   22060

Max Lipsicas
Yeshiva University
Belfer Graduate School of Sci.
2495 Amsterdam Avenue
New York, NY   10033

Farrel Lytle
The Boeing Company
P.O. Box 3999, 2T-04
Seattle, WA   98124

Archie J. McAlister
National Bureau of Standards
Matls Bldg., B150
Washington, DC   20234

J. D. E. McIntyre
Bell Lab, Rm ID-354
600 Mountain Avenue
Murray Hill, NJ   07974

Brian D. McNicol
Shell Research Ltd.
Thornton Research Centre
P.O. Box 1, CH1 3SH
Chester, England

Theodore E. Madey
National Bureau of Standards
Chem. Bldg., B248
Washington, DC   20234

Robert F. Martin
National Bureau of Standards
Matls Bldg., B348
Washington, DC   20234

Allan J. Melmed
National Bureau of Standards
Matls Bldg., B258
Washington, DC   20234

Anton Menth
Brown Boveri Research Center
5401 Baden, Switzerland

Richard P. Messmer
General Electric Corp.
Research & Development
Schenectady, NY   12301

S. R. Montgomery
W. R. Grace & Co.
Davison Chem. Div., Tech. Ctr.
5500 Chemical Road
Baltimore, MD   21226

Bruno Morosin
Sandia Labs
Albuquerque, NM   87115

S. Roy Morrison
Stanford Research Institute
Menlo Park, CA   94025

David L. Nelson
Office of Naval Research, Code 472
800 N. Quincy Street
Arlington, VA   22217

Gayle S. Painter
Oak Ridge National Lab.
P.O. Box X
Oak Ridge, TN   37830

Robert L. Park
University of Maryland
College Park, MD   20742

Daniel T. Pierce
National Bureau of Standards
Metrology Bldg., B214
Washington, DC   20234

V. Ponec
Gorlaeus Lab
University of Leiden
Wassenaarseweg 76
Leiden, Netherlands, PB 75

G. T. Pott
Koninklijke/Shell-Laboratorium
Badhuisweg 3
Amsterdam-Noord, The Netherlands·

C. J. Powell
National Bureau of Standards
Metrology Bldg., B214
Washington, DC 20234

A. K. Price
Allied Chemical
Columbia Road
Morristown, NJ 07960

Eldon B. Priestley
RCA Laboratories
Route 1 & Washington Road
Princeton, NJ 08540

Curt W. Reimann
National Bureau of Standards
Matls Bldg., B354
Washington, DC 20234

R. J. Reucroft
University of Kentucky
Dept. of Met Eng & Matls Sci
Lexington, KY 40506

Thor Rhodin
Cornell University
Applied Physics
Ithaca, NY 14853

Dana H. Ridgley
Hooker Chem. & Plastics
Long Road
Grand Island, NY 14072

William M. Riggs
Physical Electronics Industries
6509 Flying Cloud Drive
Edina, MN 55343

Ralph Roberts
The Mitre Corp.
1820 Dolley Madison Blvd.
McLean, VA 22101

Carl J. Russo
Mass. Institute of Tech.
Dept. of Matls Sci. & Engr.
Cambridge, MA 02139

Gordon A. Sargent
University of Kentucky
Lexington, KY 40506

Edward Siegel
Public Service Electric & Gas Co.
Energy Laboratory
200 Boyden Avenue
Maplewood, NJ 07040

John H. Sinfelt
Exxon Research and Eng. Co.
P.O. Box 45
Linden, NJ 07036

John R. Smith
General Motors
510 Henley
Birmingham, MI 48008

Paul Soven
University of Pennsylvania
Philadelphia, PA 19101

Fred E. Stafford
National Science Foundation
Solid State Chem.
1800 G Street, NW
Washington, DC 20550

Richard J. Stein
National Bureau of Standards
Metrology Bldg., B214
Washington, DC 20234

F. Dee Stevenson
ERDA
Div. Phys. Research
Washington, DC 20545

Paul Stonehart
United Technologies
P.O. Box 611
Middletown, CT 06457

John A. Strozier, Jr.
Brookhaven National Lab
Dept of Physics
Upton, Long Island, NY 11973

Gary W. Stupian
The Aerospace Corporation
P.O. Box 92957
Los Angeles, CA 90009

Lydon J. Swartzendruber
National Bureau of Standards
Matls Bldg., B150
Washington, DC 20234

Edward Taborek
Dept. of the Army
AMXFB-CF
Ft. Belvoir, VA 22060

Howard G. Tennent
Hercules Inc. Res. Ctr.
Market Street
Wilmington, DE 19899

John Turkevich
Princeton University
109 Rollingmead
Princeton, NJ 08540

Jens Ulstrup
Technical Univ. of Denmark
Chem. Dept. A, Bldg 207
Lyngby, Denmark 2800

George A. Uriano
National Bureau of Standards
Matls Bldg., B354
Washington, DC 20234

Alan E. Van Til
UOP Inc.
10 UOP Plaza
Des Plaines, IL 60016

Thomas H. Vanderspurt
Celanese Research Company
P.O. Box 1000 - Morris Court
Summit, NJ 07901

Theodore Vorburger
National Bureau of Standards
Metrology Bldg., B214
Washington, DC 20234

Bernard Waclawski
National Bureau of Standards
Metrology Bldg., B214
Washington, DC 20234

Grayson Walker
Dept. of the Army
AMXFB-CF
Ft. Belvoir, VA 22060

R. E. Watson
Brookhaven National Labs
Upton, NY 11719

Wendell Williams
University of Illinois
Physics Department
Urbana, IL 61801

Thomas Wolfram
University of Missouri
Physics Bldg, 223
Columbia, MO 65201

Chiang Yuan Yang
Dept. of Matls Sci. & Engr.
Mass Institute of Tech
Cambridge, MA 02139

John T. Yates
National Bureau of Standards
Chem Bldg., B268
Washington, DC 20234

Author Index
(CITED IN TEXT)

A
Adams, D.L. - 97
Aldag, Jr., A.W. - 135
Ambler, E. - 1
Anderson, A.B. - 134, 166
Appelbaum, J.R. - 165, 183, 184, 185, 186
Ashley, C.A. - 34
Azaroff, L.V. - 34

B
Ban, L.L. - 123, 124
Baetzold, R.C. - 177
Baraff, G.A. - 186
Batra, I.- 165
Bernasek, S.L. - 102
Besocke, K. - 103
Bennett, L.H. - 51, 68, 88
Blakely, M. - 178
Boudart, M. - 5, 16, 20, 46, 49, 81, 88, 98, 135, 136, 193
Breiter, MW. - 81
Brill, R. - 98
Brucker, C. - 166, 167
Brongersma, H.H. - 74

C
Callaway, J. - 176
Clementi, E. - 193
Citrin, P. - 3, 43
Ciraci, S. - 165
Cohen, M.L. - 179, 184
Coulson, C.A. - 164
Czyzewski, J.J. -50

D
Demuth, J.E. - 19
Deuss, J. - 159
Doniach, S. - 34
Domke, M. - 97, 111
Dorgelo, G.J.H. - 71, 73
Dowden, D.A. - 47, 72
Duke, C. - 43, 44, 88, 127, 167, 193, 194

K
Kasemo, B. - 46
Katzer, J. - 30, 31, 48, 51
Kjollerstrom, B. - 158
Kline, M. - 43
Knor, Z. - 95
Kunz, A.B. - 155

L
Lagally, M.G. - 69
Levine, J. - 183
Lu, K.E. - 102
Lytle, F.W. - 30, 33, 34, 42, 44

M
Madey T. - 50
Mack, R.E. - 177
Matthiess, L. - 35
Meijering, J.L. - 71
Messmer, R.P. - 87, 88, 164, 165, 171, 179
Melmed, A.J. - 91, 127, 131, 165
Moss, R.L. - 81
Müller, E.W. - 93
Mulliken, R.S. - 159

Mc
McAlister, A.J. - 55
McAllister, J. - 21, 97, 112, 119

P
Painter, G.S. - 136
Pandey, K.C. - 184, 185
Park, R.L. - 33, 43, 46, 49, 51, 69
Pendry, J.B. - 34, 44
Phillips, J.C. - 184
Ponec, V. - 47, 68, 88, 69, 70, 71, 93

R
Rodin, T. - 32, 33, 68, 166, 167
Rowe, J.E. - 129, 184, 185, 186
Russo, C.J. - 81
Rye, R.R. - 102

Subject Index

A
absorptive bonding - 19
active centres - 94, 100
appearance potential spectroscopy - 32
Anderson - Grimley formalism - 99
Anderson - Hoffman (AH) - 175
Anderson model Hamitlonian - 155, 159, 164
Auger Spectroscopy - 20, 32, 33, 42, 49, 56, 72, 73, 81, 98, 127, 135, 185
Average-t-matrix (ATA)-84

B
benzene hydrogenation - 16
biological molecules - 34
Bloch functions - 157
Brønsted Law - 95
bridged bond - 18, 156
bulk electronic properties - 3, 56, 83, 87, 88, 155

C
CaF_2 structure - 85
chemisorptive bonding - 19
chemisorptive titration - 56, 68, 73
chemisorptive luminescence - 46
cherry model - 69, 72
clusters - 69, 81, 88, 157, 164, 165
CNDO/2 - 44
coherent potential approximation - 69, 73
coordination complexes - 34
corehole excitations - 43
cyclohexane dehydrogenation - 56, 57

D
dangling bonds - 96, 99, 128, 165
demanding reactions - 20, 95
double zeta function (DZ) - 175

E
electron microscopy - 69
electron loss spectroscopy - 32
electron spin resonance (ESR) - 16, 46
electron stimulated desorption ion angular distribution (ESDIAD) - 50
enzymes - 94, 96
ethane hydrogenolysis - 56, 57
evaporated metal films - 71
extended Hückel model (EH) - 174, 175, 177
extended x-ray absorption fine structure (ESAFS) - 30, 34, 42, 43, 50

F
facile reactions - 20, 94, 100
Fermi level - 83, 139, 140
field-ion microscopy (FIM) - 127, 131

G
geometric effects - 91, 137
Green's functions - 155, 156, 158, 164

H
Hartree-Fock-approximation - 83
Hartree potential - 183
HF-SCF-LCAO calculation - 171
Hellmann-Faynman theorm - 166
high z poisons - 44
high z substrates - 43
homopolar semiconductors - 128
hydrides - 82, 83, 88
hydrogen chemisorption - 57
hyperfine fields - 51
Hückel model - 44, 164

I
infrared spectroscopy (IR) - 17, 18, 47, 48, 184
ion-neutralization spectroscopy - 32, 184, 185

K
kinetic transient method - 48
Korringa - Kohn - Rostoker (KKR) - 83, 174

L
Langmuirian surface science - 6, 7, 9, 11
Langmuir-Hinshelwood reaction - 8
LCAO - 138, 157
LEED - 20, 50, 99, 100, 101, 120, 127, 128, 187
ligand factor - 9, 135
low temperature field evaporation - 131
low z substrates - 43

M
metal clusters - 81
mid-cracking - 68
modulated molecular beam method - 20, 48, 102
Mössbauer effect - 51
muffin-tin - 83, 85, 137, 138, 174, 184
multiplet theory - 94

Material Index

A
acetylene - 167, 168, 169, 170
Ag-81
Ag-Pd-47
alumina - 45, 81, 133
Au - 34, 36, 69, 124
Au-Pt - 125

B
benzene - 56

C
C - 43, 48, 42, 193
C_2H_2 - 19
C_2H_4 - 19, 20
C_2H_6 - 19
cabosil (SiO_2) - 35
CO - 47, 49, 57, 68, 98, 99, 103, 116, 135, 142, 148, 149, 166, 177
CO_2 - 49, 100, 115, 135
Cu - 57, 73, 174
CuH - 83
Cu-Ni - 47, 56, 68, 69, 70, 71, 74
Cu-Ni-hydrides - 71
Cu-Ru - 57
Cr - 133
cyclopropane - 47

F
Fe - 133, 138, 141, 144, 166, 167, 169, 170
Fe catalysis - 97, 98, 99
formic acid - 135

G
gases - 34
glassy materials - 34

H
H_2 - 43, 46, 57, 68, 72, 87, 102, 117, 118, 135, 183, 184
H_2/D_2 - 102
hexane - 68

I
Ir - 34, 35, 131

Lewis acid - 121
Li - 171, 174, 182

M
methane - 58
Mg - 46
MgO - 16, 45
Mo - 87

N
N_2 - 21, 49, 97, 120, 121
N_2O - 96, 108
NH_3 - 21, 49, 97, 99, 119, 120, 121
Na - 193
Ni - 47, 57, 73, 97, 99, 129, 133, 135, 142, 148, 174, 175, 176, 177
NiO - 129

O
O_2 - 43, 46, 100, 101, 186
Os - 57

P
Pd - 18, 81, 83, 85, 99, 100, 101, 115, 116, 117, 123, 174
PdH - 83
Pt - 21, 34, 35, 47, 69, 81, 102, 118, 125, 126, 131, 133, 174, 177
$PtCl_2$ - 35

R
Rancy Ni - 88
Rh - 52

S
Si - 179, 183, 184, 186
SiO - 36
Sn - 133

T
Ta - 34, 35
Ti - 133
TiH - 85

V
V - 140, 141, 146

W
W - 21, 97, 19, 120, 121, 131, 175
W_2N - 119, 120, 121
WS_2 - 18
wurtzite - 128

Y
YH - 85

Z
zincblende - 128
Zeise's salt - 20
ZnO - 17

☆U.S. GOVERNMENT PRINTING OFFICE: 1977—240-848/87

NBS-114A (REV. 7-73)

U.S. DEPT. OF COMM. BIBLIOGRAPHIC DATA SHEET	1. PUBLICATION OR REPORT NO. NBS SP 475	2. Gov't Accession No.	3. Recipient's Accession No.
4. TITLE AND SUBTITLE The Electron Factor in Catalysis on Metals Proceedings of a workshop held at the National Bureau of Standards, Gaithersburg, Md., December 8-9, 1975			5. Publication Date April 1977
			6. Performing Organization Code
7. AUTHOR(S) L. H. Bennett, Editor			8. Performing Organ. Report No.
9. PERFORMING ORGANIZATION NAME AND ADDRESS NATIONAL BUREAU OF STANDARDS DEPARTMENT OF COMMERCE WASHINGTON, D.C. 20234			10. Project/Task/Work Unit No.
			11. Contract/Grant No.
12. Sponsoring Organization Name and Complete Address (Street, City, State, ZIP) National Bureau of Standards, Washington, D. C. 20234 National Science Foundation, Washington, D. C. 20550 Energy Research & Development Administration, Washington, D. C. 20545			13. Type of Report & Period Covered Final
			14. Sponsoring Agency Code

15. SUPPLEMENTARY NOTES

Library of Congress Catalog Card Number: 77-600009

16. ABSTRACT (A 200-word or less factual summary of most significant information. If document includes a significant bibliography or literature survey, mention it here.)

This book presents the proceedings of a Workshop on the Electron Factor in Catalysis on Metals held at the National Bureau of Standards, Gaithersburg, Maryland, on December 8-9, 1975. The Workshop was sponsored by the Institute for Materials Research, NBS, the Division of Materials Research of the National Science Foundation, and the Division of Conservation Research and Technology of the Energy Research and Development Administration. The purpose of the Workshop was to review the most recent experimental and theoretical investigations on catalysis on metals and related topics, and to bring together chemists, chemical engineers, surface scientists, and solid state physicists and chemists involved in research related to this topic. These proceedings summarize the four panel sessions into which the Workshop was organized: Experimental Techniques, Effect of Alloying, Geometrical Effects, and Electronic Structure.

17. KEY WORDS (six to twelve entries; alphabetical order; capitalize only the first letter of the first key word unless a proper name; separated by semicolons)

Catalysis; Characterization; Chemisorption; Electronic factor; Geometric factor; Metals; Surfaces.

18. AVAILABILITY [X] Unlimited		19. SECURITY CLASS (THIS REPORT) UNCLASSIFIED	21. NO. OF PAGES 217
[ ] For Official Distribution. Do Not Release to NTIS			
[X] Order From Sup. of Doc., U.S. Government Printing Office Washington, D.C. 20402, SD Cat. No. C13.10:475		20. SECURITY CLASS (THIS PAGE) UNCLASSIFIED	22. Price $2.80
[ ] Order From National Technical Information Service (NTIS) Springfield, Virginia 22151			

USCOMM-DC 29042-P74

There's a new look to...

... the monthly magazine of the National Bureau of Standards. Still featured are special articles of general interest on current topics such as consumer product safety and building technology. In addition, new sections are designed to ... PROVIDE SCIENTISTS with illustrated discussions of recent technical developments and work in progress ... INFORM INDUSTRIAL MANAGERS of technology transfer activities in Federal and private labs... DESCRIBE TO MANUFACTURERS advances in the field of voluntary and mandatory standards. The new DIMENSIONS/NBS also carries complete listings of upcoming conferences to be held at NBS and reports on all the latest NBS publications, with information on how to order. Finally, each issue carries a page of News Briefs, aimed at keeping scientist and consumer alike up to date on major developments at the Nation's physical sciences and measurement laboratory.

-------------------------------- (please detach here) --------------------------------